Praise for Albert-László Barabási's *Linked*

"A pleasure to read. . . . It's the fact that all of these networks can be explained and understood using the same concepts, and the same mathematics, that makes this book so important."
—*The Christian Science Monitor* (A Best Book of the Year)

"The observations of Albert-László Barabási about networks have broad applications in business. . . . Well written . . . an intellectual detective story." —*The New York Times*

"*Linked* should be mandatory reading for academics as a primer in good writing. Barabási may be a scientist, but he didn't neglect his liberal arts education; his Renaissance man's curiosity roves across history, economics, medicine, and pop culture. He writes in understandable lay-speak glittering with wit." —*The Boston Globe*

"*Linked* gets really interesting, showing how this new science promises to change the way we conduct everything from medical treatment to the war on terrorism." —*The Washington Monthly*

"All researchers dream of making a discovery that will transform their field. Albert-László Barabási can go one better. In just three years, his discovery has started making waves in fields as diverse as ecology, molecular biology, computer science, and quantum physics." —*New Scientist*

"A lively look at networks through time." —*The Washington Post*

"A sweeping look at a new and exciting science."
—Donald Kennedy, editor-in-chief, *Science Magazine*

"These laws of networks may prove as robust and universal as Newton's laws of motion."

—*strategy + business* (A Best Business Book of the Year)

"*Linked* is the best choice for the layperson, because [Barabási] minimizes the math and writes elegantly." —*Detroit Free Press*

LINKED

How Everything Is Connected to Everything Else and What It Means for Business, Science, and Everyday Life

ALBERT-LÁSZLÓ BARABÁSI

BASIC BOOKS
A Member of the Perseus Books Group
New York

Published by Basic Books, A Member of the Perseus Books Group in 2014.
First published by Perseus Publishing in 2002.

LCCN: 2014937841

ISBN (paperback): 978-0-465-08573-6
ISBN (e-book): 978-0-465-03861-9

Perseus Publishing ISBN (hardcover): 0-7382-0667-9
Penguin ISBN (paperback): 978-0-452-28439-5

10 9 8 7 6 5 4 3 2 1

Contents

THE FIRST LINK

Introduction

FEBRUARY 7, 2000, SHOULD HAVE BEEN a big day for Yahoo. Instead of the few million customers that daily flock to the Internet search engine, billions tried to enter the site. Such exploding popularity should have turned the company into the most valuable asset of the new economy. There was a problem, however. They all arrived at the exact same time and not one of them asked for a stock quote or a pecan pie recipe. Rather, they all sent, in scripted computer language, the message "Yes, I heard you!" Yahoo, as far as it could tell, had said nothing. Nevertheless, hundreds of computers in Yahoo's Santa Clara, California, headquarters were kept busy responding to these screaming ghosts, while millions of legitimate customers, who wanted a movie title or an airline ticket, waited. I was one of them. Naturally I had no idea that Yahoo was frantically busy serving ten billion ghosts. I was patient for about three minutes before I moved to a more responsive search engine. The next day the royals of the Web, Amazon.com, eBay, CNN.com, ETrade, and Excite, fell under the same spell: They too were obliged to serve billions of ghosts making the same fruitless inquiry that had handicapped Yahoo. True consumers, with shiny credit cards ready for purchases, were forced to wait on the sidelines.

Of course, getting billions of real computer users to type "Yahoo.com" into their browser at precisely 10:20 Pacific Standard Time is impossible. There are simply not enough computers around.

Early news reports construed the shutdown of the leading e-commerce sites to be the work of a group of sophisticated hackers. The consensus was that these renegade geeks, fascinated by the challenge of outsmarting sophisticated security systems, had hijacked hundreds of computers in schools, research labs, and businesses and turned them into zombies, telling Yahoo thousands of times, "Yes, I heard you." Every second, huge amounts of data were thrown at this prominent Website, much more than it could ever handle. The massive denial-of-service attack Yahoo was experiencing set off a much-publicized international hunt for the hackers responsible.

Surprisingly, the high-profile operation of the Federal Bureau of Investigation did not lead to the much-anticipated cyberterrorist organization. Instead, the FBI arrived at the suburban home of a Canadian teenager. Investigators eavesdropping on an Internet chat room overheard the teen soliciting suggestions for new targets to attack. He was caught bragging.

Hiding behind the pseudonym *MafiaBoy*, this fifteen-year-old successfully halted the operations of billion-dollar companies with access to the best computer security experts in the world. Was he a contemporary David who, armed with the humblest of home computer slingshots, beat the mega-Goliaths of the information age? In hindsight, experts agree on one thing: The attacks were not the work of a genius. They were executed using tools available to anybody on various hacker Websites. MafiaBoy's online antics revealed him to be a rank amateur, whose sloppy trail led the police right to his parents' door. In fact, his actions were more reminiscent of a Goliath than David: Lacking the know-how to penetrate any of the sites he attacked and clumsy and slow on his feet, he only managed to take down easy targets, obviously vulnerable computers from universities and small companies, which he simply instructed to bombard Yahoo with messages.

One can imagine a fifteen-year-old boy behind his bedroom door, in the glow of his computer, finding sweet satisfaction in the protracted "Yes, I heard you!" hurled at Yahoo. He must have screamed that phrase himself a million times when Mom or Dad called him to come to dinner or take out the trash. The attack succeeded with brute force, a lot of nerve, and little sophistication. But this is exactly what makes us wonder, how could this teenager's actions take out the largest corporations of the new economy? If a mere youth can wreak havoc on the Internet, what could a small group of trained and skilled professionals achieve? How vulnerable are we to such attacks?

1.

The early Christians were nothing more than a renegade Jewish sect. Regarded as eccentric and problematic, they were persecuted by both Jewish and Roman authorities. There is no historical evidence that their spiritual leader, Jesus of Nazareth, ever intended to have an impact beyond Judaism. His ideas were difficult and controversial enough for Jews, and reaching the gentiles seemed particularly hopeless. As a starter, those non-Jews who wanted to follow in his footsteps had to undergo circumcision, had to obey the laws of contemporary Judaism, and were excluded from the Temple—the spiritual center of early Jewish Christianity. Very few walked the path. Indeed, reaching them with the message was almost impossible. In a fragmented and earthbound society news and ideas traveled by foot, and the distances were long. Christianity, like many other religious movements in human history, seemed doomed to oblivion. Despite the odds, close to two billion people call themselves Christian today. How did that happen? How did the unorthodox beliefs of a small and disdained Jewish sect come to form the basis of the Western world's dominant religion?

Many credit the triumph of Christianity to the message offered by the historical figure we know today as Jesus of Nazareth. Today, marketing experts would describe his message as "sticky"—it resonated and was passed down by generations while other religious movements fizzled and died. But credit for the success of Christianity in fact goes to an orthodox and pious Jew who never met Jesus. While his Hebrew name was Saul, he is better known to us by his Roman name, Paul. Paul's life mission was to *curb* Christianity. He traveled from community to community persecuting Christians because they put Jesus, condemned by the authorities as a blasphemer, on the same level as God. He used scourging, ban, and excommunication to uphold the traditions and to force the deviants to adhere to Jewish law. Nevertheless, according to historical accounts, this fierce persecutor of Christians underwent a sudden conversion in the year 34 and became the fiercest supporter of the new faith, making it possible for a small Jewish sect to become the dominant religion in the Western world for the next 2,000 years.

How did Paul's efforts succeed? He understood that for Christianity to spread beyond Judaism, the high barriers to becoming a Christian had to be abolished. Circumcision and the strict food laws had to be relaxed. He took his message to the original disciples of Jesus in Jerusalem and received the mandate to continue evangelization without demanding circumcision.

But Paul understood that this was not enough: The message had to spread. So he used his firsthand knowledge of the social network of the first century's civilized world from Rome to Jerusalem to reach and convert as many people as he could. He walked nearly 10,000 miles in the next twelve years of his life. He did not wander randomly, however; he reached out to the biggest communities of his era, to the people and places in which the faith could germinate and spread most effectively. He was the first and by far the most effective salesperson of Christianity, using theology and social networks equally effectively. So should he, or Jesus, or the message be credited for Christianity's success? Could it happen again?

2.

There are huge differences between MafiaBoy and Paul: MafiaBoy's was an act of destruction. Paul, despite his initial intentions, became a bridge builder between early Christian communities. But the two have something important in common: Both were masters of the network. Though neither of them thought about it in these terms, the key to their success was the existence of a complex network that offered an effective medium for their actions. MafiaBoy operated on a network of computers—the Internet is the fastest and most effective way to reach the largest number of people at the turn of the third millennium. Paul was a master of first-century social and religious links, the only network at the beginning of the modern era that could carry and spread a faith. Neither of them fully grasped the forces that aided them in their actions. But nearly 2,000 years after Paul we are making the first inroads toward understanding what made Paul and MafiaBoy successful. We now know that the answer lies as much in the structure and topology of the networks on which they operated as in their ability to navigate them.

Paul and MafiaBoy succeeded because we are all connected. Our biological existence, social world, economy, and religious traditions tell a compelling story of interrelatedness. As the great Argentinean author Jorge Luis Borges put it, "everything touches everything."

3.

"There be dragons there!" wrote the ancient mapmakers, marking off the frightening unknown. As adventurous explorers penetrated every region of the globe, these monster-marked patches gradually disappeared. But there are still lots of dragon-infested areas in our mental map of how the different parts of the world fit together, from the microscopic universe locked within a cell to the unbounded world of the Internet. The good news is that recently scientists have been learning to map our interconnectivity. Their maps are shedding new light on our weblike universe, offering surprises and challenges that could not even be imagined a few years ago. Detailed maps of the Internet have unmasked the Internet's vulnerability to hackers. Maps of companies connected by trade or ownership have traced the trail of power and money in Silicon Valley. Maps of interactions between species in ecosystems have offered glimpses of humanity's destructive impact on the environment. Maps of genes working together in a cell have provided insights into how cancer works. But the real surprise has come from placing these maps side by side. Just as diverse humans share skeletons that are almost indistinguishable, we have learned that these diverse maps follow a common blueprint. A string of recent breathtaking discoveries has forced us to acknowledge that amazingly simple and far-reaching natural laws govern the structure and evolution of all the complex networks that surround us.

4.

Have you ever seen a child take apart a favorite toy? Did you then see the little one cry after realizing he could not put all the pieces back together again? Well, here is a secret that never makes the headlines: We have taken apart the universe and have no idea how to put it back together. After spending trillions of research dollars to disassemble nature in the last century, we are just now acknowledging that we have no clue how to continue—except to take it apart further.

Reductionism was the driving force behind much of the twentieth century's scientific research. To comprehend nature, it tells us, we first must decipher its components. The assumption is that once we understand the parts, it will be easy to grasp the whole. Divide and conquer; the devil is in the details. Therefore, for decades we have been forced to see the world through its constituents. We have been trained to study atoms and superstrings to understand the universe; molecules to comprehend life; individual genes to understand complex human behavior; prophets to see the origins of fads and religions.

Now we are close to knowing just about everything there is to know about the pieces. But we are as far as we have ever been from understanding nature as a whole. Indeed, the reassembly turned out to be much harder than scientists anticipated. The reason is simple: Riding reductionism, we run into the hard wall of complexity. We have learned that nature is not a well-designed puzzle with only one way to put it back together. In complex systems the components can fit in so many different ways that it would take billions of years for us to try them all. Yet nature assembles the pieces with a grace and precision honed over millions of years. It does so by exploiting the all-encompassing laws of self-organization, whose roots are still largely a mystery to us.

Today we increasingly recognize that nothing happens in isolation. Most events and phenomena are connected, caused by, and interacting with a huge number of other pieces of a complex universal puzzle. We have come to see that we live in a small world, where everything is linked to everything else. We are witnessing a revolution in the making as scientists from all different disciplines discover that complexity has a strict architecture. We have come to grasp the importance of networks.

With the Internet dominating our life, the word *network* is on everybody's lips these days, featured in company names and popular journal titles. After September 11, witnessing the deadly power of terrorist networks, we had to get used to yet another meaning of the term. Very few people realize, however, that the rapidly unfolding science of networks is uncovering phenomena that are far more exciting and revealing than the casual use of the word *network* could ever convey. Some of these discoveries are so fresh that many of the key results still circulate as unpublished papers within the scientific community. They open up a novel perspective on the interconnected world around us, indicating that networks will dominate the new century to a much greater degree than most people are yet ready to acknowledge. They will drive the fundamental questions that form our view of the world in the coming era.

This book has a simple aim: to get you to think networks. It is about how networks emerge, what they look like, and how they evolve. It shows you a Web-based view of nature, society, and business, a new framework for understanding issues ranging from democracy on the Web to the vulnerability of the Internet and the spread of deadly viruses.

Networks are present everywhere. All we need is an eye for them. As you move from link to link within this book, you will learn to see society as a complex social network and to grasp the smallness of this great world in which we live. You will come to understand how and why Paul succeeded and how, despite some obvious differences, his social milieu was similar to the one we experience today. You will see the challenges doctors face when they attempt to cure a disease by focusing on a single molecule or gene, disregarding the complex interconnectedness of living matter. You will be reminded that MafiaBoy is not alone in attacking networks. You will come to appreciate how the Internet, often viewed as entirely human in its creation, has become more akin to an organism or an ecosystem, demonstrating the power of the basic laws that govern all networks. You will see how the emergence of terrorism is also ruled by the laws of network formation and how these deadly webs take advantage of the fundamental robustness of nature's webs. You'll wonder at the amazing similarities among such diverse systems as the economy, the cell, and the Internet, using one to grasp the other. This will be an eye-opening trip across disciplines that I hope will challenge you to step out of the box of reductionism and explore, link by link, the next scientific revolution: the new science of networks.

THE SECOND LINK

The Random Universe

ON SEPTEMBER 18, 1783, IN ST. PETERSBURG Leonhard Euler started the day as usual. He gave a mathematics lesson to one of his grandchildren and took up some calculations on the flight of balloons. Just three months earlier, south of Lyon, the Montgolfier brothers had launched an enormous balloon that rose 6,500 feet into the air and landed safely about a mile away. Euler was working out the mechanics of the balloon's motion as the Montgolfier brothers were preparing to launch a sheep into the air in front of King Louis XVI in Paris, a flight that took place the next day, on September 19. Euler never heard about the event, however. After lunch, working with his assistants, he made some calculations on the orbit of the recently discovered planet Uranus. The equations introduced by him, capturing the planet's peculiar orbit, would lead decades later to the discovery of the planet Neptune. Euler did not live to witness that discovery either. About five o'clock in the afternoon, he suffered a brain hemorrhage and uttered, "I am dying," before losing consciousness. He died that evening, ending the most prolific career in mathematics of all time.

Euler, a Swiss born mathematician who spent his career in Berlin and St. Petersburg, had an extraordinary influence on all areas of mathematics, physics, and engineering. Not only was the importance of his discoveries unparalleled, their sheer quantity is also overwhelming. *Opera Omnia*, the still incomplete record of Euler's collected works, currently runs to over seventy-three volumes, six hundred pages each. The last seventeen years of Euler's life, between his return to St. Petersburg in 1766 and his death at the age of 76, were rather tumultuous. Yet, despite many personal tragedies, about half of his works were written during these years. These include a 775-page treatise on the motion of the moon, an influential algebra textbook, and a three-volume discussion of integral calculus, completed while he continued to publish an average of one mathematics paper per week in the journal of the St. Petersburg Academy. The amazing thing is that he barely wrote or read a single line during this time. Having partially lost his sight soon after returning to St. Petersburg in 1766, Euler was left completely blind after a failed cataract operation in 1771. The thousands of pages of theorems were all dictated from memory.

Three decades earlier, his sight intact, Euler had written a short paper addressing an amusing problem that originated in Königsberg, a town not too far from Euler's home in St. Petersburg. Königsberg, a flowering city in eastern Prussia, did not suspect in the early eighteenth century the sad and war-torn fate that awaited it as host for one of the fiercest battles of the Second World War. Contemporary etchings show a thriving city on the banks of the Pregel, where a busy fleet of ships and their trade offered a comfortable life to the local merchants and their families. The healthy economy allowed city officials to build not fewer than seven bridges across the river. Most of these connected the elegant island Kneiphof, which was caught between the two branches of the Pregel, with other parts of the city. Two additional bridges crossed the two branches of the river (Figure 2.1). The people of Königsberg, enjoying a time of peace and prosperity, amused themselves with mind puzzles, one of which was: "Can one walk across the seven bridges and never cross the same one twice?" No one was to find such a path until a new bridge was built in 1875.

Almost 150 years before the new bridge, in 1736, Euler offered a rigorous mathematical proof stating that with the seven bridges such a path does not exist. He not only solved the Königsberg problem but in his brief paper inadvertently started an immense branch of mathematics known as graph theory. Today graph theory is the basis for our thinking about networks. During the centuries after Euler it grew into a

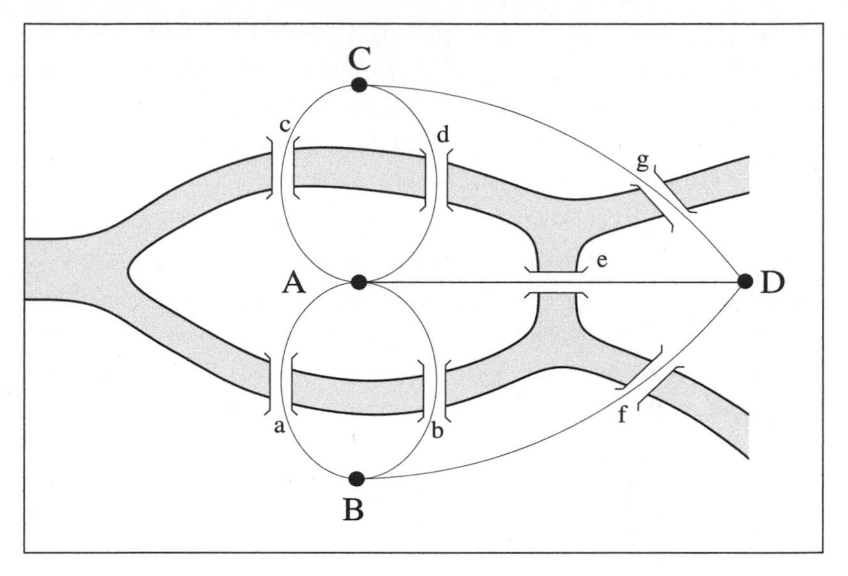

Figure 2.1 Königsberg Bridges. *The layout of Königsberg before 1875, with Kneiphof island (*A*) and the land area* D *caught between the two branches of the Pregel River. Solving the Königsberg problem meant finding a route around the city that would require a person to cross each bridge only once. In 1736, Leonhard Euler gave birth to graph theory by replacing each of the four land areas with nodes (*A *to* D*) and each bridge with a link (*a *to* g*), obtaining a graph with four nodes and seven links. He then proved that on the Königsberg graph, a route crossing each link only once does not exist.*

mature field, to which most great mathematicians contributed. To open the door on the field of networks, let us briefly revisit the reasoning process that led Euler to the introduction of the first graph.

1.

Euler's proof is simple and elegant, easily understood even by those not trained in mathematics. Nevertheless, it is not the proof that made history but rather the intermediate step that he took to solve the problem. Euler's great insight lay in viewing Königsberg's bridges as a *graph*, a collection of *nodes* connected by *links*. For this he used nodes to represent each of the four land areas separated by the river, distinguishing them with letters *A*, *B*, *C*, and *D*. Next he called the bridges the links and connected with lines those pieces of land that had a bridge between them. He thus obtained a graph whose nodes were pieces of land and links were bridges.

Euler's proof that in Königsberg there is no path crossing all seven bridges only once was based on a simple observation. Nodes with an odd number of links must be either the starting or the end point of the journey. A continuous path that goes through all bridges can have only one starting and one end point. Thus, such a path cannot exist on a graph that has more than two nodes with an odd number of links. As the Königsberg graph had four such nodes, one could not find the desired path.

For our purpose the most important aspect of Euler's proof is that the existence of the path does not depend on our ingenuity to find it. Rather, *it is a property of the graph.* Given the layout of the Königsberg bridges, no matter how smart we are, we will never succeed at finding the desired path. The people of Königsberg finally agreed with Euler, gave up their fruitless search, and in 1875 built a new bridge between B and C, increasing the number of links of these two nodes to four. Now only two nodes (A and D) with an odd number of links remained. It was then rather straightforward to find the desired path. Perhaps the creation of this path was the hidden rationale behind building the bridge?

In retrospect, Euler's unintended message is very simple: Graphs or networks have properties, hidden in their construction, that limit or enhance our ability to do things with them. For more than two centuries the layout of Königsberg's graph limited its citizens' ability to solve their coffeehouse problem. But a change in the layout, the addition of only one extra link, suddenly removed this constraint.

In many ways Euler's result symbolizes an important message of this book: The construction and structure of graphs or networks is the key to understanding the complex world around us. Small changes in the topology, affecting only a few of the nodes or links, can open up hidden doors, allowing new possibilities to emerge.

Graph theory boomed after Euler with contributions made by mathematical giants such as Cauchy, Hamilton, Cayley, Kirchhoff, and

Pólya. They uncovered just about everything that is known about large but ordered graphs, such as the lattice formed by atoms in a crystal or the hexagonal lattice made by bees in a beehive. Until the mid-twentieth century the goal of graph theory was simple: It aimed to discover and catalogue the properties of the various graphs. Famous problems included finding a way to escape from a maze or labyrinth, first solved in 1873, or finding a sequence of moves with a knight on a chess board such that each square is visited only once and the knight returns to its starting point. Some of the more difficult problems have gone unsolved for centuries.

Two centuries passed after Euler's inspiring work before mathematicians moved from studying the properties of various graphs to asking the quintessential question of how graphs, or, more commonly, networks, came about. Indeed, how do real networks form? What are the laws governing their appearance and structure? These questions, and the first answer, did not come until the 1950s, when two Hungarian mathematicians made a revolution in graph theory.

2.

One afternoon in late 1920s Budapest, a seventeen-year-old youth cantered with a weird gait through the streets and stopped in front of an elegant shoe shop that sold custom-made shoes. With his strangely shaped feet, on which normal shoes would never fit, he could indeed use a cobbler. But new shoes were not the occasion of this visit. After knocking on the store's door—an act that would have seemed just as odd back then as today—he entered, ignoring the saleswoman at the counter, and went up to a fourteen-year-old boy in the back of the shop.

"Give me a four digit number," he said.

"2,532," came the wide-eyed boy's reply as he stared at the strange creature. The older boy did not let him stare too long, however.

"The square of it is 6,441,024," he continued. "Sorry, I am getting old and I cannot tell you the cube. How many proofs of the Pythagorean theorem do you know?"

"One," replied the youngster.

"I know thirty-seven," and without taking a breath he continued. "Did you know that the points of a straight line do not form a countable set?" After showing the sharp boy Cantor's proof as evidence, his business at the cobbler's store finished, he said, "I must run," and so he did, turning on his heel and galloping out of the store.

Paul Erdős galloped on to become the presiding genius and most famous misfit of the twentieth century. He wrote more than 1,500 mathematics papers before his death in 1996. This output, unparalleled since Euler, contained eight articles published with another Hungarian mathematician, Alfréd Rényi. These eight papers addressed for the first time in history the most fundamental question pertaining to our understanding of our interconnected universe: How do networks form? Their solution laid the foundation of the theory of random networks. This elegant theory so profoundly determined our thinking about networks that we are still struggling to break away from its hold.

3.

Organize a party for a hundred guests who have been selected and invited because they do not know a single other person on the guest list. Offer this group of strangers wine and cheese, and they will immediately start to chat, as human beings' inborn desire to meet and know each other inevitably brings them together. Soon you will see thirty to forty groups of two or three. Now mention to one guest that the red wine in the unlabeled dark green bottles is a rare twenty-year-old vintage port, far better than that with the red label. But ask that guest to share this information only with his or her new acquaintances. You know that your expensive port is fairly safe, because your friend has only had time to meet two or three people in the room. However, guests will inevitably become bored talking to the same person for too long and move on to join other groups. An outside observer would not notice anything special. Yet there are invisible social links between people who met earlier but now belong to different groups. As a consequence, subtle paths start connecting people who are still strangers to each other. For example, though John has not met Mary yet, they have both met Mike, and so

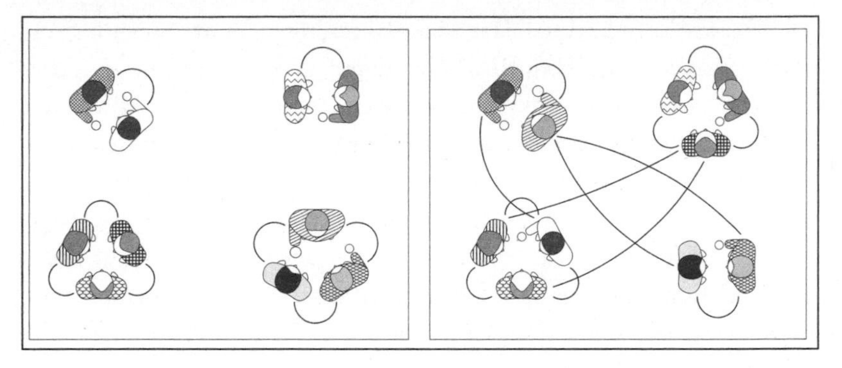

Figure 2.2 The Party. *At a party with ten guests, none of whom initially knows one another, social ties form as the guests start chatting in small groups. At first, the groups are isolated from each other (left panel). Indeed, though there are social links (shown as continuous lines) between those in the same group, everyone outside of that group is still a stranger. As time goes on (right panel), three guests move to different groups and a giant cluster emerges. Although not everyone knows everyone else, there is now a single social network that includes all the guests. By following the social links, one can now find a path between any two guests.*

there is a path from John to Mary through Mike. If John knew about the wine, chances are that now Mary knows too, since she could hear it from Mike, who was told by John. As time goes on, the guests will be increasingly interwoven by such intangible links, creating a fine web of acquaintances that includes a sizable portion of the guests. The expensive wine is increasingly endangered as its identity is passed from a tiny group of insiders to more and more chatting groups (Figure 2.2).

Assuming that each person passes on the information to all of her or his new acquaintances, will the reputation of the fine port reach all of the guests before the end of the party? To be sure, if all were to get to know each other, everybody would be pouring the superior wine from the unlabeled bottle. But even if each encounter took only ten minutes, meeting all ninety-nine others would take about sixteen hours. Parties rarely last that long. Thus, you might feel that you could reveal the identity of the wine to your friend and reasonably hope that some would be left at the end of the party.

Paul Erdős and Alfréd Rényi begged to differ. "A mathematician is a machine that turns coffee into theorems," Erdős used to say, quoting Rényi. A particularly lucky cup of coffee turned into a much quoted theorem: If each person gets to know at least *one* other guest, then soon everybody will be drinking the reserve port. According to Erdős and Rényi, it would take only thirty minutes to form a single invisible social web that includes all guests in the room. Minutes after you hear the recommendation for the wine, you may find yourself tipping an empty bottle into your expectant glass.

4.

The guests we met at the cocktail party are part of a problem in graph theory, the branch of mathematics pioneered by Euler. The guests are the nodes, and every encounter creates a social link between them. Thus a web of acquaintances—a graph—emerges, a bunch of nodes connected by links. Computers linked by phone lines, molecules in our body linked by biochemical reactions, companies and consumers linked by trade, nerve cells connected by axons, islands connected by bridges are all examples of graphs. Whatever the identity and the nature of the nodes and links, for a mathematician they form the same animal: a *graph* or a *network*.

Despite its elegance, simplifying all webs into graphs poses some formidable challenges. While society, the Internet, a cell, or the brain can all be represented by graphs, each is clearly very different from the others. It is hard to imagine much commonality between human society, where we make friends and acquaintances through a combination of random encounters and conscious decisions, and the cell, where the unforgiving laws of chemistry and physics govern all reactions between molecules. There must be a clear difference in the rules that govern the placement of links in the various networks we encounter in nature. Finding a model to describe all of these different systems seems, on its face, an insurmountable challenge.

Yet the ultimate goal of all scientists is to find the simplest possible explanation for very complex phenomena. Erdős and Rényi took on this challenge by proposing an elegant mathematical answer to describe all complex graphs within a single framework. Since different systems follow such disparate rules in building their own networks, Erdős and Rényi deliberately disregarded this diversity and came up with the simplest solution nature *could* follow: connect the nodes randomly. They decided that the *simplest* way to create a network was to play dice: Choose two nodes and, if you roll a six, place a link between them. For any other roll of the dice, do not connect these two nodes but choose a different pair and start over. Therefore, Erdős and Rényi viewed graphs and the world they represented as fundamentally random.

"There is an old debate," Erdős liked to say, "about whether you create mathematics or just discover it. In other words, are the truths already there, even if we don't yet know them?" Erdős had a clear answer to this question: Mathematical truths are there among the list of absolute truths, and we just rediscover them. Random graph theory, so elegant and simple, seemed to him to belong to the eternal truths. Yet today we know that random networks played little role in assembling our universe. Instead, nature resorted to a few fundamental laws, which will be revealed in the coming chapters. Erdős himself created mathematical truths and an alternative view of our world by developing random graph theory. Not privy to nature's laws in creating the brain and society, Erdős hazarded his best guess in assuming that God enjoys playing dice. His friend Albert Einstein, at Princeton, was convinced of the opposite: "God does not play dice with the universe."

5.

Let's go back to our cocktail party and the exercise in random graph theory. You start with a large number of isolated nodes. Then you randomly add links between the nodes, mimicking the random encounters between the guests. If you add only a few connections, the only consequence of your activity will be that some of the nodes will pair up. If you continue adding links, you will inevitably connect some of these pairs to each other, forming clusters of several nodes. But when you add enough links such that each node has an average of *one* link, a miracle happens: A unique giant cluster emerges. That is, most nodes will be part of a single cluster such that, starting from any node, we can get to any other by navigating along the links between the nodes. This is the moment when your expensive wine is in danger, since a rumor can reach everyone who belongs to the giant cluster. Mathematicians call this phenomenon the emergence of a giant component, one that includes a large fraction of all nodes. Physicists call it percolation and will tell you that we just witnessed a phase transition, similar to the moment in which water freezes. Sociologists would tell you that your subjects had just formed a community. Though different disciplines may have different terminology, they all agree that when we randomly pick and connect pairs of nodes together in a network, something special happens: The network, after placing a critical number of links, drastically changes. *Before*, we have a bunch of tiny isolated clusters of nodes, disparate groups of people that communicate only within the clusters. *After*, we have a giant cluster, joined by almost everybody.

6.

Each of us is part of a large cluster, the worldwide social net, from which no one is left out. We do not know everybody on this globe, but it is guaranteed that there is a path between any two of us in this web of people. Likewise, there is a path between any two neurons in our brain, between any two companies in the world, between any two chemicals in our body. Nothing is excluded from this highly interconnected web of life. Paul Erdős and Alfréd Rényi told us why: It requires *only one link per node* to stay connected. One acquaintance per person, one link to at least one other neuron for each neuron in the brain, the ability to participate in at least one reaction for each chemical in our body, trade with at least one other company in the business world. One is the threshold. If nodes have less than one connection on average, then our network breaks into tiny noncommunicating clusters. If there is more than one connection per node, that danger becomes remote.

Nature repeatedly and extravagantly exceeds the one-link minimum. Sociologists estimate that we know between 200 and 5,000 people by name. An average neuron is connected to dozens of others, some to thousands. Each company is inevitably linked to hundreds of suppliers and customers; some of the biggest have links to millions. In our body, most molecules take part in far more than a single reaction—some, like water, in hundreds. Thus, real networks not only are connected but are well beyond the threshold of one. Random network theory tells us that as the average number of links per node increases beyond the critical one, the number of nodes left out of the giant cluster decreases exponentially. That is, the more links we add, the harder it is to find a node that remains isolated. Nature does not take risks by staying close to the threshold. It well surpasses it. Consequently, the networks around us are not just webs. They are very dense networks from which nothing can escape and within which every node is navigable. This is why there are no islands of people completely isolated from society at large and why all molecules in our body are integrated into a single complex cellular map. This is why the Apostle Paul's message reached people he never met and why MafiaBoy made headlines: Along the links their actions easily affected millions.

7.

Erdős and Rényi's discovery of this very special moment when a giant cluster emerges through a phase or percolation transition was a huge event in graph theory, but not because it made the unbelievable prediction that only one acquaintance is required to form a society. Rather, it was largely because, before Erdős and Rényi, graph theory had not dealt with cocktail parties, social networks, or random graphs. It focused almost exclusively on regular graphs, which contain no ambiguity about their structure. But when it comes to such complex systems as the Internet or the cell, regular graphs are the exception rather than the norm. Erdős and Rényi acknowledged for the first time that real graphs, from social networks to phone lines, are not nice and regular. They are hopelessly complicated. Humbled by their complexity, the two assumed that these networks are random.

In retrospect, it is not surprising that this unlikely pair of mathematicians were the ones to turn around a respectable field of mathematics by injecting randomness into it. Chance and randomness were very much a part of their lives. Though Rényi was seven years younger than Erdős, they knew each other thanks to the friendship between their parents back in Budapest. By the time they started working together, after meeting up in Amsterdam in 1948, both had lived through rather tumultuous times. Subject to the *Numerus Clausus* laws that limited the number of Jews admitted to university, Rényi had worked in a shipyard after high school. After winning a math and Greek competition, he was allowed to enter the university in 1939. Soon after finishing his mathematical studies he was called to forced labor, from which he somehow escaped.

Erdős and his colleagues, who were familiar with Rényi's resistance activities during the war, deeply admired and respected him. Rényi had boldly disguised himself in the uniform of the Hungarian fascists, Nyilas, to help his friends escape the concentration camps. According to one story, Rényi entered the Budapest ghetto dressed as a Nyilas soldier and managed to escort his parents out. He also lived for years in Nazi-controlled Budapest using false documents. Only those aware of the realities of the Nazi terror could truly appreciate the courage needed to perform these acts. Not surprisingly, Rényi's ability to focus on mathematics was highly constrained until the end of the war, when in 1946 he traveled to Leningrad to continue his studies. There his creativity exploded. He not only learned and absorbed number theory in record time, despite his limited Russian language skills, but also proved some fundamental theorems on one of the notoriously difficult problems of number theory, the Goldbach conjecture. Thus, when he met Erdős two years later in Amsterdam, he was no longer the aspiring young mathematician and family friend but a well-known scientist with an international reputation.

Erdős by then had already developed his trademark traveling-mathematician lifestyle. He would show up at his colleagues' doorsteps and proclaim, "My brain is open," an invitation to join in his tireless pursuit of mathematical truth. His only permanent job offer came from the University of Notre Dame, in South Bend, Indiana. Arnold Ross, at that time the chairman of the math department, offered Erdős a visiting professorship on very generous terms: He could come and go as he pleased, since he had an assistant who would pick up the lectures where he left them off.

A Catholic liberal arts college, Notre Dame was not the prominent university it would become decades later. Nevertheless it offered Erdős a quiet and comfortable work environment and the opportunity for frequent discussions with his priest colleagues, which Erdős, with his unique perspective on the universe and deity, particularly enjoyed. Once asked about his time there, he remarked tongue-in-cheek, "There are too many plus signs," a reference to the numerous crucifixes about campus. When Notre Dame eventually offered to turn Erdős's status into a permanent one, on the same comfortable terms, Erdős politely refused. Perhaps losing the randomness and unpredictability that had characterized his life was too much for him to fathom.

8.

The Amsterdam meeting between Erdős and Rényi was the start of a very close friendship and collaboration that resulted in over thirty joint publications before Rényi's early death at the age of forty-nine in 1970. Among these publications were the eight legendary papers on graph theory. The first, published more than a decade after the Amsterdam meeting, addressed for the first time the important questions of how graphs form. Their use of randomness to tackle graph theory problems is most evident when we look at how many links nodes have in a graph or network. Regular graphs are unique in that each node has *exactly* the same number of links. Indeed, in a two-dimensional mesh of perpendicular lines forming a simple square lattice each node has exactly four links, or in a hexagonal lattice of a beehive each node is connected to exactly three others.

Such regularity is clearly absent from random graphs. The premise of the random network model is deeply egalitarian: We place the links completely randomly; thus all nodes have the same chance of getting one—just as in Las Vegas, where supposedly we all have the same chance of hitting the jackpot. At the end of the day, however, only a few of our fellow gamblers walk away richer. Similarly, if we place the links randomly in a graph, some nodes will get more links than others. Some might even have bad luck and get nothing for a while. The random world of Erdős and Rényi can be simultaneously unfair and generous: It can make some poor and others rich. Yet a far-reaching prediction of Erdős and Rényi's theory tells us that this only appears to be so. If the network is large, despite the links' completely random placement, almost *all nodes will have approximately the same number of links*.

One way to see this is to interview all guests as they leave the cocktail party, asking them how many acquaintances they made. When everybody leaves, we can draw a histogram by plotting how many of the guests have one, two, or exactly *k* new acquaintances. For the random network model of Erdős and Rényi the shape of the histogram was derived and proved exactly in 1982 by one of Erdős's students, Béla Bollobás, professor of mathematics at the University of Memphis in the United States and Trinity College in the United Kingdom. The result shows that the histogram follows a Poisson distribution, which has some unique properties that will follow us throughout this book. A Poisson distribution has a prominent peak, indicating that the majority of nodes have the same number of links as the average node does. On the two sides of the peak the distribution rapidly diminishes, making significant deviations from the average extremely rare.

Translated back to a society of 6 billion people, a Poisson distribution tells us that most of us have roughly the same number of friends and acquaintances. It predicts that it is exponentially rare to find someone who deviates from the average by having considerably more or fewer links than the average person. Therefore, random graph theory predicts that if we assign social links randomly, we end up with an extremely democratic society, where all of us are average and very few deviate from the norm to be extremely social or utterly asocial types. We obtain a network with a very uniform fabric in which the mean is the norm.

Erdős and Rényi's random universe is dominated by averages. It predicts that most people have roughly the same number of acquaintances; most neurons connect roughly to the same number of other neurons; most companies trade with roughly the same number of other companies; most Websites are visited by roughly the same number of visitors. As nature blindly throws the links around, in the long run no node is favored or singled out.

9.

The random network theory of Erdős and Rényi has dominated scientific thinking about networks since its introduction in 1959. It created several paradigms that are consciously or unconsciously imprinted on the minds of everyone who deals with networks. It equated complexity with randomness. If a network was too complex to be captured in simple terms, it urged us to describe it as random. Sure enough, society, the cell, communication networks, and the economy are all complex enough to fit the bill.

You may be thinking that there is something fishy about this random universe, in which all nodes are equal. Would I be able to write this book if the molecules in my body decided to react to each other randomly? Would there be nations, states, schools, and churches or any other manifestations of social order if people interacted with each other completely randomly? Would we have an economy if companies selected their consumers randomly, replacing their salespeople with millions of dice? Most of us *feel* that we do not live in such a random world—that there has to be some order behind these complex systems.

Why, then, would two such unparalleled intellects as Erdős and Rényi choose to model the emergence of networks as a completely random process? The answer is simple: They never planned to provide a universal theory of network formation. They were far more intrigued by the mathematical beauty of random networks than by the model's ability to faithfully capture the webs nature created around them. To be sure, in their seminal 1959 paper they did mention that "the evolution of graphs may be considered as a rather simplified model of the evolution of certain communication nets (railway, road or electric network systems, etc.)." But, despite this brief journey into the real world, their work in this area was motivated by a deep curiosity about the mathematical depths of the problem rather than by its applications.

Erdős would be the first to agree with us that real networks must have organizing principles that distinguish them from the random network model they introduced in 1959. But for him this would be beside the point. By using the hypothesis of randomness he opened a window to a new world, whose mathematical beauty and consistency was the main driving force behind the subsequent work in graph theory.

Until recently we had no alternative for describing our interlinked universe. Thus random networks came to dominate our ideas on network modeling. Complex real networks were viewed as fundamentally random.

Erdős holds the record for suggesting good problems and making sure that somebody else solved them. Though he never owned more than a few clothes that fit into a small leather suitcase that he always traveled with, he often offered monetary rewards for solutions or proofs to problems that he found interesting—$5 for a problem he considered simple, $500 for a truly difficult one. And he would happily pay if the proof was delivered. Never mind that often a $1 problem turned out to be more difficult than a $500 one. The lucky mathematicians who earned one of his rewards never cashed his checks anyway. Most of them framed them. The reward was a unique recognition by the presiding genius of the century; no cash amount could match its spiritual value.

Let us follow Erdős's example and ask a question he left untouched. What do *real* networks look like? Posing a problem in such a sloppy way would never have satisfied him. It is too broad. It may not even have a unique answer. And most likely we can never offer a rigorous proof. Thus it could not possibly be from the Transfinite Book, the ultimate depository in Erdős's world of all good mathematical proofs and theorems. But though the question might not have won his approval, in the coming chapters, we will see that it makes a huge difference outside the world of mathematics.

THE THIRD LINK

Six Degrees of Separation

IN 1912, JUST AS ANNA ERDŐS DISCOVERED she was pregnant with her third child, Paul, the streets of Budapest were abuzz with talk about a new collection of poems and prose by the best Hungarian and international writers. The first edition had sold out before the literary critics could even get to it, and the second printing was also disappearing when the first serious reviews appeared in newspapers around the country. By then Anna Erdős had entered the hospital, given birth to Paul, and gone home, only to discover that her two older daughters were the victims of a scarlet fever epidemic that was tearing through Budapest.

Despite the city's many personal tragedies, enthusiasm for the new literary phenomenon was unabating. The book's popularity was rooted in a minor detail: All the poems and short stories were fake. In *Igy irtok ti*, or *This Is How You Write*, Frigyes Karinthy, a twenty-five-year-old virtually unknown poet and writer, invented what he called *literary caricature*. The volume is a collection of poems and short stories that appear to be written by a who's who of world literature. If you were familiar with the authors, you could easily recognize their styles. Each piece is a cunning parody that, like a distorting mirror, keeps the mimicked author recognizable while changing all the proportions. Karinthy applied his vitriolic and annihilating humor with equal ease on deceased giants and close friends. And his arrow was often deadly: The authors he most venomously parodied are known to us only through his book; their actual works are lost in the unforgiving sink of literary taste and history.

Igy irtok ti is one of the most read books in Hungarian history. It made Karinthy an instant celebrity. Never again did he have to wait for the bus in the bus station—he simply waved to it from wherever he was, and the drivers, with wide smiles, stopped for him. He wrote most of the time behind the expansive glass windows of the Central Café in the heart of Budapest. Passersby often performed a strange dance. As they walked by the window, they suddenly stopped, turned, and peered through the window at the working writer, as if he were an exotic species in a new aquarium.

Almost two decades after *Igy irtok ti*, in 1929, at about the same time that the seventeen-year-old Erdős was lecturing about the Pythagorean theorem in the shoe store a few streets away from the Central Café, Karinthy published his forty-sixth book, *Minden masképpen van (Everything Is Different)*, a collection of fifty-two short stories. By now he was recognized as the genius of Hungarian literature. Everyone, however, was still waiting for "The Book," the novel that would define Karinthy and guarantee his place among literature's immortals. The critics openly voiced concern that Karinthy was selling out his unique talent by writing short stories that drew quick bucks. Karinthy, whose incredibly disordered and chaotic life was spent between coffeehouses and a hectic and noisy home, failed to deliver the long awaited tome. The short story collection was a critical failure and soon sank into obscurity. It has been out of print ever since. I have visited most bookstores and antiquaries in Budapest and cannot find a trace of it. But there is one story, entitled "Láncszemek," or "Chains," that deserves our attention.

"To demonstrate that people on Earth today are much closer than ever, a member of the group suggested a test. He offered a bet that we could name any person among earth's one and a half billion inhabitants and through at *most five* acquaintances, one of which he knew personally, he could link to the chosen one," writes Karinthy in "Láncszemek." And indeed, Karinthy's fictional character immediately links a Nobel prizewinner to himself, noting that the Nobelist must know King Gustav, the Swedish monarch who hands out the Nobel prize, who in turn is a consummate tennis player and plays occasionally with a tennis champion who happens to be a good friend of Karinthy's character. Remarking that linking to celebrities is easy, Karinthy's character demands a more difficult assignment. Next he tries to link a worker in Ford's factory to himself: "The worker knows the manager in the shop, who knows Ford; Ford is on friendly terms with the general director of Hearst Publications, who last year became good friends with Árpád Pásztor, someone I not only know, but is to the best of my knowledge a good friend of mine—so I could easily ask him to send a telegram via the general director telling Ford that he should talk to the manager and have the worker in the shop quickly hammer together a car for me, as I happen to need one." Though these short stories have been neglected, Karinthy's 1929 insight that people are linked by at most five links was the first published appearance of the concept we know today as "six degrees of separation."

1.

Six degrees was rediscovered almost three decades later, in 1967, by Stanley Milgram, a Harvard professor who turned the concept into a much celebrated, groundbreaking study on our interconnectivity. Amazingly, Milgram's first paper on the subject occasionally reads like an English translation of Karinthy's "Láncszemek" rewritten for an audience of sociologists. Milgram, perhaps the most creative practitioner of experimental psychology, is best known for a series of highly debated experiments probing the conflict between obedience to authority and personal conscience. But his intellect was wide-ranging, and he soon became interested in the structure of our social network, a topic that was frequently discussed by sociologists at Harvard and MIT during the late sixties.

Milgram's goal was to find the "distance" between any two people in the United States. The question driving the experiment was, how many acquaintances would it take to connect two randomly selected individuals? To get started, he first chose two target persons, the wife of a divinity graduate student in Sharon, Massachusetts, and a stock broker in Boston. He picked Wichita, Kansas, and Omaha, Nebraska, as starting points for the study because "from Cambridge, these cities seem vaguely 'out there,' on the Great Plains or somewhere." There was little consensus about how many links it would take to connect people from these remote areas. Milgram himself pointed out in 1969, "Recently I asked a person of intelligence how many steps he thought it would take, and he said that it would require 100 intermediate persons, or more, to move from Nebraska to Sharon."

Milgram's experiment entailed sending letters to randomly chosen residents of Wichita and Omaha asking them to participate in a study of social contact in American society. The letter contained a short summary of the study's purpose, a photograph, and the name and address of and other information about one of the target persons, along with the following four-step instructions:

HOW TO TAKE PART IN THIS STUDY

1. ADD YOUR NAME TO THE ROSTER AT THE BOTTOM OF THIS SHEET, so that the next person who receives this letter will know who it came from.

2. DETACH ONE POSTCARD. FILL IT OUT AND RETURN IT TO HARVARD UNIVERSITY. No stamp is needed. The postcard is very important. It allows us to keep track of the progress of the folder as it moves toward the target person.

3. IF YOU KNOW THE TARGET PERSON ON A PERSONAL BASIS, MAIL THIS FOLDER DIRECTLY TO HIM (HER). Do this only if you have previously met the target person and know each other on a first name basis.

4. IF YOU DO NOT KNOW THE TARGET PERSON ON A PERSONAL BASIS, DO NOT TRY TO CONTACT HIM DIRECTLY. INSTEAD, MAIL THIS FOLDER (POSTCARDS AND ALL) TO A PERSONAL ACQUAINTANCE WHO IS MORE LIKELY THAN YOU TO KNOW THE TARGET PERSON. You may send the folder

> to a friend, relative or acquaintance, but it must be someone you know on a first name basis.

Milgram had a pressing concern: Would any of the letters make it to the target? If the number of links was indeed around one hundred, as his friend guessed, then the experiment would likely fail, since there is always someone along such a long chain who does not cooperate. It was therefore a pleasant surprise when within a few days the first letter arrived, passing through only two intermediate links! This would turn out to be the shortest path ever recorded, but eventually 42 of the 160 letters made it back, some requiring close to a dozen intermediates. These completed chains allowed Milgram to determine the number of people required to get the letter to the target. He found that the median number of intermediate persons was 5.5, a very small number indeed—and coincidentally, amazingly close to Karinthy's suggestion. Round it up to 6, however, and you get the famous "six degrees of separation."

As Thomas Blass, a social psychologist who has devoted the last fifteen years to in-depth research on the life and work of Stanley Milgram, pointed out to me, Milgram himself never used the phrase "six degrees of separation." John Guare originated the term in his brilliant 1991 play of that title. After an extremely successful season on Broadway, the play was made into a movie with the same title. In the play, Ousa (played by Stockard Channing in the movie), musing about our interconnectedness, tells her daughter, "Everybody on this planet is separated by only six other people. Six degrees of separation. Between us and everybody else on this planet. The president of the United States. A gondolier in Venice. . . . It's not just the big names. It's anyone. A native in a rain forest. A Tierra del Fuegan. An Eskimo. I am bound to everyone on this planet by a trail of six people. It's a profound thought. . . . How every person is a new door opening up into other worlds."

Milgram's study was confined to the United States, linking people "out there" in Wichita and Omaha to "over here" in Boston. For Guare's Ousa, however, six degrees applied to the whole world. Thus a myth was born. Because more people watch movies than read sociology papers, Guare's version has prevailed in popular thought.

Six degrees of separation is intriguing because it suggests that, despite our society's enormous size, it can easily be navigated by following social links from one person to another—a network of *six billion* nodes in which any pair of nodes are on average *six* links from each other. Perhaps we should be surprised that there *is* a path between any two people. Yet we saw in the previous chapter that being connected requires very little—barely more than one social link per person. As we all have many more than one link, each of us is a part of the giant network that we call society.

Stanley Milgram awakened us to the fact that not only are we connected, but we live in a world in which no one is more than a few handshakes from anyone else. That is, we live in a *small world*. Our world is small because society is a very dense web. We have far more friends than the critical one needed to keep us connected. Yet is six degrees something uniquely human, tied somehow to our desire to form social links? Or do other kinds of networks look the same? Answers to these questions surfaced only a few years ago. We now know that social networks are not the only small worlds.

2.

"Suppose all the information stored on computers everywhere were linked. . . . All the best information in every computer at CERN and on the planet would be available to me and anyone else. There would be a single global information space." This was the dream of Tim Berners-Lee in 1980 while working as a programmer at the European Organization for Nuclear Research, commonly known by its French acronym, CERN, in Geneva, Switzerland. To turn his dream into reality, he wrote a program that allowed computers to share information—to link to each other. By inventing the links, Berners-Lee released a genie whose existence had been unknown to us. In less than ten years the genie turned into the World Wide Web, one of the largest ever human-made networks. It is a virtual network whose nodes are Webpages that have it all: news, movies, gossip, maps, pictures, recipes, biographies, and books. If it can be written, drawn, or photographed, chances are there is already a node on the Web containing it in some form.

The power of the Web is in the links, the uniform resource locators (URLs) that allow us to move with the click of a mouse from one page to another. They allow us to surf, locate, and string together information. These links turn the collection of individual documents into a huge network spun together by mouse clicks. They are the stitches that keep the fabric of our modern information society together. Remove the links, and the genie would spectacularly vanish. Huge inaccessible databases would be left behind, the contemporary ruins of an interconnected world.

How large is the Web today? How many Web documents and links are out there? Until recently no one knew for sure—there's no single organization to keep track of all the nodes and links. It was Steve Lawrence and Lee Giles, working at the NEC Research Institute at Princeton, who took up this unique challenge in 1998. Their measurements indicated that in 1999 the Web had close to a billion documents—not bad for a virtual society born less than a decade earlier. Considering that it grows much faster than human society, chances are that by the time this book is published there will be more Web documents than people on Earth.

But the real issue isn't the overall size of the Web. It's the distance between any two documents. How many clicks does it take to get from the home page of a high-school student in Omaha to the Webpage of a Boston stockbroker? Despite the billion nodes, could the Web be a "small world"? The answer to this question is not irrelevant to anybody who surfs the Web. If Webpages are thousands of clicks from each other, it is hopeless to find any document without a search engine. Finding that the Web was not a small world would also indicate that the networks behind society and the online universe were fundamentally different. If that were the case, to fully understand networks we would need to understand why and how this difference emerges. Therefore, at the end of 1998 I set out with Réka Albert, a Ph.D. student, and Hawoong Jeong, a postdoctoral associate—both working at that time in my research group at the physics department at the University of Notre Dame—to grasp the size of the world behind the Web.

Our first goal was to obtain a map of the Web, essentially an inventory of all Webpages and the links connecting them. The information contained in such a map would be truly unparalleled. If we were to construct a similar map for society, it would have to include each person's professional and personal interests and chart everyone she or he knew. It would make Milgram's experiment seem clumsy and obsolete by allowing us to find, in seconds, the shortest path to any person in the world. It would be a must-use tool for everyone from politicians to salespeople and epidemiologists. Of course, such a social search engine is impossible to build, since it would take at least a lifetime to interrogate all 6 billion people on the earth to learn about their friends and acquaintances. Yet there is something magical about the Web that sets it apart from society: We can navigate its links instantaneously. It is just a matter of clicks.

Unlike our current society, the Web is digital. This allows us to write a piece of software that downloads any document, finds all the links on it, then visits and downloads the documents to which they point, continuing until all pages on the Web are captured. If you let such a program loose, in theory it will return a complete map of the Web. In the computer world, this software is called a *robot* or *crawler* because it crawls through the Web without human supervision. The big search engines, such as Alta Vista or Google, have thousands of computers running numerous robots that constantly look for new documents on the Web. Our little research group clearly could not compete on their scale. So Jeong created a robot to accomplish a more modest goal. First it gave us a map of the nd.edu domain by mapping about 300,000 documents within the University of Notre Dame, a rather eclectic collection containing everything from philosophy course Web pages to Irish music fan sites. But we were not concerned about the content of the pages. We were interested only in the links that told us how to travel from one page to another. With such a map at hand, we could then measure the distance between any two pages within the university.

Just as Milgram saw some of his letters reaching the target person in two steps while others took as many as eleven, our results indicated lots of variability in the distances between Web documents. For example, my graduate students have links to my Webpage; thus they are one click away from me. Yet going from my Webpage to the homepage of a philosophy major would often require twenty clicks. What was astonishing, however, is that, taken together, these paths were not as long as the vastness of the Web would suggest. The measurements indicated that pages were on average eleven clicks away from each other. Paraphrasing Guare's title, we could say there are eleven degrees of separation at Notre Dame.

However, the Webpages within our university, the nd.edu domain, represent only a tiny subset of the World Wide Web. The full Web in 1999 was at least 3,000 times larger. Would this mean that the distance between two randomly selected nodes on the World Wide Web was also 3,000 times longer than the eleven clicks our measurements indicated? That is, would it take a full 33,000 clicks to get from one page to another on the Web? To answer this question we needed a map of the full Web. The problem was that nobody had one. Even the largest search engines that tirelessly scan the Web with thousands of computers have managed to cover less than 15 percent of the Web's full size. Could we determine the separation for the full Web without such a map? The answer was yes. But we had to use a method commonly employed in statistical mechanics—the field of physics that regularly deals with random systems with unpredictable components or outcomes.

Our approach had a simple premise: If the Web is too large to fit into our computer, then we should study many smaller pieces of it that do fit. For example, we took a small portion of the Web, with only 1,000 nodes, and calculated the separation between any two nodes on this tiny sample. Next we took a slightly larger piece, with 10,000 nodes, and determined the separation again. We repeated this for the largest systems our computer allowed us to use and looked for trends in the obtained node-to-node distances. The results indicated that the average separation between the nodes increased much more slowly than the number of documents, following a very simple and reproducible expression.[1] This finding allowed us to predict the separation on the

---

1. We found the separation to be proportional to the logarithm of the number of nodes in the network. That is, if we denote d to be the average separation between the nodes on a Web of N Webpages, then this separation followed the equation $d = 0.35 + 2\log N$, where $\log N$ denotes the base-10-logarithm of N.

full Web as long as we know the total number of documents out there. That number was provided by the NEC group. They estimated the size of the publicly indexable Web to be around 800 million nodes at the end of 1998. Thus our expression predicted that the diameter of the Web was 18.59, close to 19. As Guare might say: nineteen degrees of separation. While surfing might give you a different impression, in reality the Web is a small world. Any document is on average only nineteen clicks away from any other.

3.

Taken together, Milgram's six degrees and the Web's nineteen degrees suggest that behind the short observed distances there is something more fundamental than humanity's desire to spread social links all over the globe. This suspicion was confirmed by subsequent discoveries which demonstrated that small separations are common in just about every network scientists have had a chance to study. Indeed, species in food webs appear to be on average two links away from each other; molecules in the cell are separated on average by three chemical reactions; scientists in different fields of science are separated by four to six coauthorship links; and the neurons in the brain of the C. *elegans* worm are separated by fourteen synapses. In fact, it appears that the Web holds the absolute record at nineteen degrees, as all other networks studied so far display a separation between two and fourteen.

Nineteen degrees may appear to be drastically far from six degrees. This is not the case, however. What is important is that huge networks, with hundreds of millions or billions of nodes, collapse, displaying separation far shorter than the number of nodes they have. Our society, a network of six billion nodes, has a separation of six. The Web, with close to a billion nodes, has a separation of nineteen. The Internet, a network of hundreds of thousands of routers, has a separation of ten. Seen from this perspective, the difference between six and nineteen is negligible.

The natural question is: Why? How do networks achieve such a uniformly short path despite consisting of billions of nodes? The answer lies in the highly interconnected nature of these networks. In the previous chapter, we saw that random networks require only one link per node to form a giant cluster. The question is, what if, as usually happens in real networks, nodes have many more links than that? At the critical point when the average connectivity is around one per node, the separation between nodes could be rather large. But as we add more links, the distance between the nodes suddenly collapses. Consider a network in which the nodes have on average k links. This means that from a typical node we can reach k other nodes with one step. There are, however, k^2 nodes two links away and roughly k^d nodes exactly d links away. Therefore, if k is large, for even small values of d the number of nodes you can reach can become very large. Within a few steps you have reached all nodes to be found, which explains why the average separation is so short in most networks.

These arguments can be easily turned into a mathematical formula that predicts the separation in a random network as a function of the number of nodes.[2] The origin of the small separation is a logarithmic term present in the formula. Indeed, the logarithm of even a very large number is rather small. The ten-based logarithm of a billion is only nine. For example, if we have two networks, both with an average of ten links per node, but one 100 times larger than the other, the separation of the larger net will be only two degrees higher than the separation of the smaller one. The logarithm shrinks the huge networks, creating the small worlds around us.

4.

One of the most absentminded people of his generation, Karinthy was well-known for forgetting meetings he had arranged ahead of time. Dezső Kosztolányi, Karinthy's close friend and literary rival, once remarked, "I have got to run home because Karinthy promised that he would visit us, and perhaps he forgot that he promised, and he will indeed come." Interestingly, six degrees appears to follow a very Karinthyan path: forgotten,

2. If we have N nodes in the network, k^d must not exceed N. Thus, using $k^d = N$, we obtain a simple formula that works well for random networks, telling us that the average separation follows the equation $d = \log N/\log k$.

reformulated, and rediscovered in the popular press and scientific texts alike. I have no idea who originally discovered the six degrees concept. The earliest written account that I know of comes from Karinthy. But how did he get it? Did he think of it by himself? In view of his unparalleled intellect and fondness for unexpected and unconventional ideas, it is not inconceivable. Or did he hear about it from others in the coffeehouse, as his short story suggests? We will perhaps never have an answer. But it is interesting to speculate on the subsequent turn of events.

Karinthy's short story was published in 1929, when Erdős, also living in Budapest, was seventeen years old. As even unsuccessful books of Karinthy's were literary events, it is not unlikely that Erdős read or heard of the "Chains" story, in which Karinthy postulates that all people on the earth can be connected by a chain of five acquaintances. The same conjecture could even be made about Alfred Rényi, who, though only nine years old when "Chains" appeared, had a unique affinity for literature. Indeed, he was known to have been good friends with many writers, including Karinthy's son, Ferenc, a well-known writer himself.

Erdős teamed up with Alfred Rényi in 1959 to write their famous string of eight papers on random networks. The papers do contain the expression giving the network's diameter as a function of the number of nodes. Should either of them have cared, they could have easily shown that Karinthy's intuition was correct, since the many social links we have shrink even gigantic webs into truly tiny worlds. They never mention this application in their papers, however, and we will probably never know if they amused themselves with the idea while taking breaks between proofs and theorems. But the links do not stop there. Stanley Milgram published his experiments uncovering the 5.5 links in 1967, four decades after Karinthy's five-link conjecture and almost a decade after Erdős and Rényi introduced the random network theory. He did not seem to have been aware of the body of work on networks in graph theory and most likely had never heard of Erdős and Rényi. He is known to have been influenced by the work of Ithel de Sole Pool of MIT and Manfred Kochen of IBM, who circulated manuscripts about the small-world problem within a group of colleagues for decades without publishing them, because they felt they had never "broken the back of the problem." Incidentally, Milgram is a child of a Hungarian father and Romanian mother who immigrated to the United States and settled in the Bronx. Could his father or uncles, who often visited, have been aware even anecdotally of Karinthy's five degrees? Could his real interest in the problem have been rooted in stories he overheard as a child? This again is something that we will never know, but it certainly suggests some interesting paths in the evolution of the idea of six degrees.

5.

The six/nineteen degrees phrase is deeply misleading because it suggests that things are easy to find in a small world. This could not be further from the truth! Not only is the desired person or document six/nineteen links away, but so are all people or documents. In other words, six—or ten or nineteen—can either be a very small number or a very large one, depending on what you're trying to do. Since the average number of links on any given Web document is around seven, this means that while we can follow only seven links from the first page, there are 49 documents two clicks away, 343 three clicks away, and so on. By the time we reach the nodes that are exactly nineteen degrees away, in principle we would have checked 10^{16} documents, 10 million times more than the total number of pages on the Web. This contradiction has an easy resolution: Some of the links we meet along the road will point back to pages that we have seen before. Thus they are not "new" links. But even if it takes only one second to check a document, it would still take over 300 million years to get to all documents that are nineteen clicks away! Nevertheless, despite the abundance of choices, we sometimes find documents rather quickly, even without search engines.

The trick, of course, is that we do not follow all links. Rather, we use clues. Indeed, if we are looking for information on Picasso and are faced with three choices on a given Webpage, we are more apt to follow the modern art link than either the link for a famous wrestler or a frog's love life. By *interpreting* the links, we avoid having to check all the pages within nineteen degrees and can zero in on the desired page within a few clicks. While this method seems to be the most efficient, it almost always fails to find the shortest path. Indeed, it is always possible that the wrestler whose Webpage we bypassed balances his tough guy image with a link to the best Picasso site. But most people looking for Picasso would ignore the wrestler's link and eventually follow a longer path to the destination. The computer, having no taste or bias (yet), will chew through with equal excitement the wrestler, modern art, and frog's love life pages, pragmatically following the links to all of them. By trying all the possible paths, it will inevitably locate the shortest one, independent of the content of the intermediate pages.

Finding Picasso on the Web highlights a fundamental problem with six degrees: Milgram's method overestimated the shortest distance between two people in the United States. Six degrees is really an upper limit. There is an enormous number of paths with widely different lengths between any two people. Milgram's subjects were never aware of the shortest path to their target. This is like being lost in a huge maze where we can see only the corridors and doors next to us. Even if we have a compass and we know that the exit is toward the north, finding it could be woefully inefficient and time-consuming. With a map of the maze in hand, we could be out in five minutes. Similarly, Milgram's letters would have followed the shortest path between Omaha and Boston only if all participants had had a map that compiled the social links of all Americans. Lacking such a map, they forwarded the message to those that they *thought* were most likely to take it in the right direction. For example, if you wish to be introduced to the president of the United States, you would try to think of somebody who knows the president. Most likely you would settle on your senator or representative. As most of us do not know our senator on a first name basis, we would try to find somebody who does and who would be willing to broker a meeting with the president. That would take at least three handshakes. In the meantime, you might have no clue that the gentleman you sat next to a few days earlier at a dinner party went to school with the president. Thus in reality you are only two degrees away from the president. Similarly, the paths recorded by Milgram's experiment were invariably longer than the shortest possible. Thus, the real separation in society was clearly overestimated. It must be shorter than six—perhaps shorter than Karinthy's five. We don't have a social search engine, so we may never know the real number with total certainty.

6.

Six degrees is the product of our modern society—a result of our insistence on keeping in touch. It is aided by our relatively newfound ability to communicate over great distances—often over thousands of miles. The global village we've grown used to inhabiting is a new reality for humans. The ancestors of most Americans lost contact with those they left behind in the old country. From the cattle herds on the prairies or the gold mines of the Rocky Mountains it was impossible to reach loved ones separated by oceans and continents. No postcards, no phone calls. In the subtle social network of those days, it was rather difficult to activate the links that had been broken when people moved. That changed in this century as the mail system, the telephone, and then air travel demolished barriers and shrank physical distances. Today immigrants to America can choose to maintain their links to the people they leave behind. We can and do keep in touch. I keep track of my relatives and friends even if they are as far away as Korea or eastern Europe. The world has collapsed irreversibly in the twentieth century. And it is undergoing yet another implosion right now, as the Internet reaches to every corner of the world. Though we are nineteen clicks away from everybody on the Web, we are only one click away from our friends. They might have hopped three cities and five jobs since we last met in person. But no matter where they are, we can usually find them on the Internet if and when we wish to do so. The world is shrinking because social links that would have died out a hundred years ago are kept alive and can be easily activated. The number of social links an individual can actively maintain has increased dramatically, bringing down the degrees of separation. Milgram estimated six. Karinthy five. We could be much closer these days to three.

"Small worlds" are a generic property of networks in general. Short separation is not a mystery of our society or something peculiar about the Web: Most networks around us obey it. It is rooted in their structure—it simply doesn't take many links for me to reach a huge number of Webpages or friends. The resulting small worlds are rather different from the Euclidean world to which we are accustomed and in which distances are measured in miles. Our ability to reach people has less and less to do with the physical distance between us. Discovering common acquaintances with perfect strangers on worldwide trips repeatedly reminds us that some people on the other side of the planet are often closer along the social network than people living next door. Navigating this non-Euclidean world repeatedly tricks our intuition and reminds us that there is a new geometry out there that we need to master in order to make sense of the complex world around us.

THE FOURTH LINK

Small Worlds

WHEN MARK GRANOVETTER SUBMITTED his first-ever paper for publication, he was still a graduate student at Harvard, but he had high hopes for his manuscript. Harvard in the late 1960s was the right place at the right time. Networks were about to infest sociology, and Harvard and MIT were the hotbeds of the new ideas. A series of lectures by Harrison White, a pioneer of the network perspective in social sciences, exposed Granovetter to social networks early in his graduate studies. Many of the new ideas proved to have fallen on fertile ground in his doctoral thesis, which brought under a sociological microscope an issue that sooner or later plagues all graduates: how people get jobs. Instead of polishing his résumé and going to job fairs, Granovetter crossed the Charles River to Newton, Massachusetts. Whereas today Newton is a wealthy suburb of Boston, in the late sixties it was a working-class neighborhood. Aiming to find out how people "network"—use their social connections to land a new job—he interviewed dozens of managerial and professional workers, asking them who helped them find their current job. Was it a friend? He kept getting the same reply: No, it was not a friend. It was just an acquaintance. This reminded Granovetter of the classic chemistry lesson demonstrating how weak hydrogen bonds hold huge water molecules together, and that image, stuck in his mind since his freshman year, inspired his first research paper, a long, revealing manuscript on the importance of the weak social ties in our lives. He mailed it out in August 1969 to the *American Sociological Review*. In December he received word that two anonymous referees had rejected the paper. As one put it, the manuscript should not be published for "an endless series of reasons that immediately come to mind." Terribly discouraged, Granovetter did not touch the paper until three years later. In 1972 he submitted a somewhat shortened version of the manuscript to a different journal, the *American Journal of Sociology*. This time he had better luck, and the paper was finally published in May 1973, four years after its first submission. Today Granovetter's paper, *The Strength of Weak Ties*, is recognized as one of the most influential sociology papers ever written. It is one of the most cited as well, featured as a Citation Classic by *Current Contents* in 1986.

In *The Strength of Weak Ties* Granovetter proposed something that sounds preposterous at first: When it comes to finding a job, getting news, launching a restaurant, or spreading the latest fad, our weak social ties are more important than our cherished strong friendships. As he put it, the structure of the social network around an ordinary person, whom he calls Ego, is rather generic. "Ego will have a collection of close friends, most of whom are in touch with one another—a densely knit clump of social structure. Moreover, Ego will have a collection of acquaintances, few of which know each other. Each of these acquaintances, however, is likely to have close friends in his own right and therefore to be enmeshed in a closely knit clump of social structure, but one different from Ego's."

Hidden within Granovetter's argument there is an image of a society that is very different from the random universe Erdős and Rényi depicted. In his view society is structured into highly connected clusters, or close-knit circles of friends, in which everybody knows everybody else. A few external links connecting these clusters keep them from being isolated from the rest of the world. If Granovetter's description is correct, then the network describing our society has a rather peculiar structure. It is a collection of *complete graphs*, tiny clusters in which each node is connected to all other nodes within the cluster (Figure 4.1). These complete graphs are linked to each other by a few weak ties between acquaintances belonging to different circles of friends.

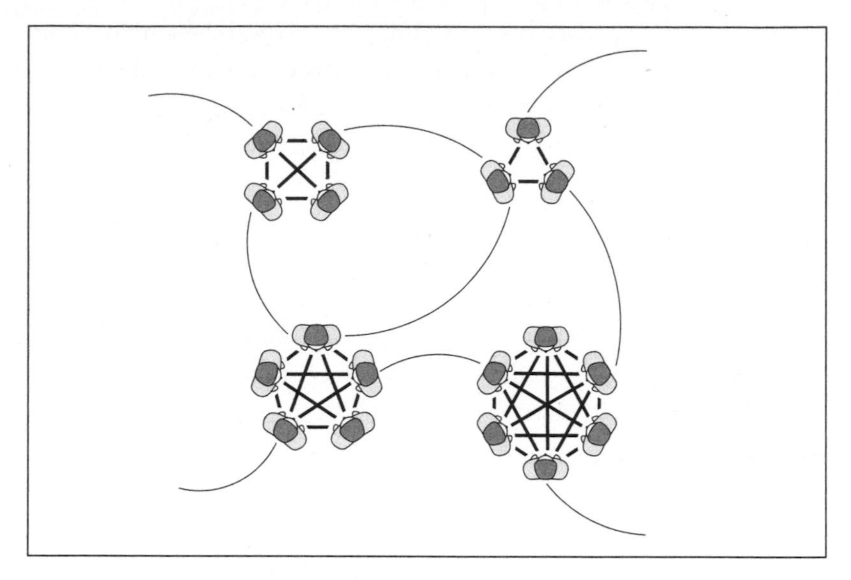

Figure 4.1 Strong and Weak Ties. *In Mark Granovetter's social world, our close friends are often friends with each other as well. The network behind such a clustered society consists of small, fully connected circles of friends connected by strong ties, shown as bold lines. Weak ties, shown as thin lines, connect the members of these friendship circles to their acquaintances, who have strong ties to their own friends. Weak ties play an important role in any number of social activities, from spreading rumors to getting a job.*

Weak ties play a crucial role in our ability to communicate with the outside world. Often our close friends can offer us little help in finding a job. They move in the same circles we do and are inevitably exposed to the same information. To get new information, we have to activate our weak ties. Indeed, managerial workers are more likely to hear about a job opening through weak ties (27.8 percent of the cases) than through strong ties (16.7 percent). The weak ties, or acquaintances, are our bridge to the outside world, since by frequenting different places they obtain their information from different sources than our immediate friends.

In a random network there would be no circle of friends, as our links to other nodes are completely random. In the Erdős-Rényi social universe the likelihood of my two closest friends knowing each other is the same as the chance that an Australian cobbler's best friend is an African tribal chief. But that is not what our society looks like. In most cases two good friends know each other's friends. They often go to the same parties, frequent the same pubs, and watch the same movies. The stronger the tie between two people, the larger the overlap between their circles of friends. Though Granovetter's argument about the importance of weak ties at first glance may seem counterintuitive and even paradoxical, it formulates a simple truth about our social organization. Granovetter's society, a fragmented web of fully connected clusters communicating through weak ties, is truer to our daily experience than the completely random picture offered by Erdős and Rényi. To fully understand the structure of society, somehow the theory of random networks had to be reconciled with the clustered reality depicted by Granovetter. It took almost three decades to accomplish this. Interestingly, the clue for a possible solution did not come from sociology or graph theory.

1.

Across from the Central Café, a few paces from Karinthy's favorite window, you descend through a small door and narrow underground staircase into one of the elite studio theaters of Budapest. Appropriately named the *Kamra*, or Closet, since it holds only about ten actors on the stage and a hundred people in the audience, the seats at its performances are highly coveted by those familiar with Budapest's burgeoning theater life. The last performance I watched in the *Kamra* did away with the curtain to save space, forcing the audience to guess exactly when the play ended. It was hard to miss, though, as suddenly everyone around me burst into tumultuous applause, which was echoed and amplified by the black walls of the small underground cavern. In no time the chaotic thunder gave way to unison clapping. Our palms came together at precisely the same moment, united by a mysterious force that urged us to clap in phase, as if following the baton of an invisible conductor. As the actors bowed, disappeared backstage, and reappeared, the rhythmic applause grew even stronger. Its synchrony dissolved temporarily as the clapping gathered speed and strength, only to reappear in full force a few seconds later.

Synchronized clapping is hardly unique to the tiny *Kamra* theater of Budapest. It is a regular occurrence after theater performances, concerts, or sports events in eastern Europe and is occasionally heard all over the world. It spontaneously emerged, for example, in Madison Square Garden when the audience unconsciously synchronized its clapping to honor Wayne Gretzky, the legendary hockey player, before his retirement from the New York Rangers in 1999. Spontaneous and mysterious, synchronized clapping offers a wonderful example of self-organization following strict laws extensively researched by physicists and mathematicians. Some species of fireflies are also subject to these laws. In southwest Asia they gather by the millions around tall mangrove trees, flashing periodically. Then suddenly, all the fireflies begin to switch their fluorescent tails on and off at exactly the same moment, turning the beacon-shaped tree into a huge pulsing light bulb visible for miles. A subtle urge to synchronize is pervasive in nature. Indeed, it drives the firing of thousands of pacemaker cells in the heart and brings into synchrony the menstrual cycles of women who live together for long periods of time.

Duncan Watts, working on his Ph.D. in applied mathematics at Cornell University in the mid-1990s, was asked to investigate a peculiar problem: how crickets synchronize their chirping. Male crickets attract females by chirping loudly. Unlike many humans, crickets eschew the spotlight by carefully listening to the other crickets around them, adjusting their chirp to match that of their neighbors. Put many of them together and from the cacophony a symphony emerges that we often enjoy on the back porch on humid summer nights.

Watts does not fit the stereotypical image of a bookish mathematician. Possessing an agile mind, he has the rare ability to stop, step back, and reflect on his work, changing direction if he needs to. Indeed, the cricket study turned him into a student of social networks and eventually a sociologist, a transformation made official in 2000 when he was offered a professorship in the department of sociology at Columbia University.

While struggling to grasp how crickets synchronize, Watts was struck by the concept of six degrees of separation, planted in his head by his father during a casual conversation. People wonder about things like six degrees all the time, but such coffeehouse philosophy rarely leads to serious research. Watts thought that to fully comprehend how crickets synchronize he needed to understand how they pay attention to each other. Do all crickets listen to every other cricket that is chirping? Or do some pick a favorite one and try to synchronize with that one only? What is the structure of the network encoding how crickets, or people, influence each other? Finding himself thinking more and more about networks, and less and less about crickets, Watts approached his Ph.D. advisor, Steven Strogatz for advice. An applied mathematics professor at Cornell with a distinguished record in the study of chaos and synchronization, Strogatz is not known for allowing an unconventional idea to pass him by. Soon they were off to uncharted territories, taking networks beyond the boundaries set by Erdős and Rényi.

Watts started his voyage into networks with a simple question: What is the likelihood that two friends of mine know each other? As we have just seen, this question has a clear answer in random network theory. Because the nodes are linked randomly, my two best friends have the same chance of knowing each other as do a gondolier from Venice and an Eskimo fisherman. Clearly, as Granovetter argued twenty-five years earlier, that is not how society works. We are part of clusters in which everybody knows everybody else. Thus my two best friends will inevitably know each other. To gather evidence about the clustered nature of society in terms that are acceptable to a mathematician or physicist we need to be able to *measure* clustering. To achieve this, Watts and Strogatz introduced a quantity called the *clustering coefficient*. Let's assume that you have four good friends. If they are all friends with *each other* as well, you can connect each of them with a link, obtaining altogether six friendship links. Chances are, however, that some of your friends are not friends with each other. Then the real count will give fewer than six links—let's say, four. In this case the clustering coefficient for your circle of friends is 0.66, obtained by dividing the number of actual links between your friends (four) by the number of links that they could have if they were all friends with each other (six).

The clustering coefficient tells you how closely knit your circle of friends is. A number close to 1.0 means that all your friends are good friends with each other. On the other hand, if the clustering coefficient is zero, then you are the only person who holds your friends together, as they do not seem to enjoy each other's company. Granovetter's vision of society includes many highly connected clusters, linked to each other by weak ties. Such a highly clustered network should have a large clustering coefficient. To obtain quantitative evidence that society is indeed full of such clusters, we would need to measure the clustering coefficient for each person on Earth. As there are no maps telling us who is connected to whom and who is friends with whom, this is an impossible task. Fortunately, however, a peculiar subset of society regularly publishes their social ties. We can therefore look for clustering among this unusual group.

2.

Today Paul Erdős is famous not only for his countless theorems and proofs, but also for a concept he inspired: the Erdős number. Erdős published over 1,500 papers with 507 coauthors. It is an unparalleled honor to be counted among his hundreds of coauthors. Short of this, it is a great distinction to be only two links from him. To keep track of their distance from Erdős, mathematicians introduced the *Erdős number*. Erdős has Erdős number zero. Those who coauthored a paper with him have Erdős number one. Those who wrote a paper with an Erdős coauthor have Erdős number two, and so on. A low Erdős number is a matter of pride—so much so that some suspect that counterfeit collaborations may have been concocted after Erdős's death in 1996 to lower someone's number. Consequently, mathematicians all around the world have been (and still are) scrambling to figure out their distance from this eccentric center of the math universe. To ease their search, Jerry Grossmann, a professor of mathematics at Oakland University in Rochester, Michigan, maintains a detailed Webpage collecting the Erdős numbers for thousands of mathematicians, allowing any published mathematician to calculate his or her own.

Most mathematicians turn out to have rather small Erdős numbers, being typically two to five steps from Erdős. But Erdős's influence goes well beyond his immediate field. Economists, physicists, and computer scientists also can be easily connected to him. Einstein has Erdős number two. Paul Samuelson, the Nobel prize–winning economist, has five. James D. Watson, the codiscoverer of the double helix, has eight. Noam Chomsky, the famous linguist, has four. Even William H. (Bill) Gates, founder of Microsoft, who has published little science, has an Erdős number of four. My Erdős number is also four: Erdős wrote a paper with Joseph E. Gillis, who had George H. Weiss among his seventeen coauthors, who in turn worked with H. Eugene Stanley, my Ph.D. advisor, with whom I have coauthored a book and over a dozen scientific articles.

The very existence of the Erdős number demonstrates that the scientific community forms a highly interconnected network in which all scientists are linked to each other through the papers they have written. The smallness of most Erdős numbers indicates that this web of science truly is a small world. As it only rarely happens that the authors of a publication do not personally know each other, coauthorships represent strong social links. Consequently, the web of science is a small-scale prototype of our social network, with the unique feature that its links are regularly published. Indeed, so that researchers can locate papers on a certain topic, all scientific publications are recorded in computerized databases; this automatically creates a detailed digital record of the social and professional links between scientists. We can therefore use them to study the structure of the collaboration network.

This is exactly what a group of us did in the spring of 2000. Tamás Vicsek, a distinguished researcher and chairman of the department of biological physics at Eötvös University in Budapest during the academic year 1999–2000, organized a year-long program focusing on biological physics at the Institute of Advanced Study, located in a charming medieval Buda castle overlooking the Danube. Zoltán Néda, a physicist from Romania, was one of the participants, and he had brought along Erzsébet Ravasz, at that time a masters student in Néda's group. Also joining the team was András Schubert, an expert on sociometrics working for the Hungarian Academy, who had access for research purposes to large coauthorship databases. Together with Vicsek, Ravasz, Néda, Schubert, and Hawoong Jeong, we linked all mathematicians through papers published between 1991 and 1998, reassembling the highly interwoven network of 70,975 mathematicians connected by over 200,000 coauthorship links. If the mathematicians had chosen their coauthors randomly, the resulting random network would be predicted by the Erdős-Rényi theory to have a very small clustering coefficient, approximately 10^{-5}. However, our measurements indicated that the clustering coefficient for the real collaboration network is about 10,000 times larger than that, proving that mathematicians do not pick their collaborators randomly. Rather, they form a highly clustered network, similar to the one spotted by Granovetter in society at large.

Unknown to us, Mark Newman, a physicist at the Santa Fe Institute, had also been investigating the collaboration graph of scientists—in particular physicists, medical doctors, and computer scientists—asking questions similar to the ones we were asking in Budapest. Newman, whose expertise ranges from random systems to species extinctions in ecosystems, recognized the unique opportunity our computerized world offers us to finally understand networks. Before turning to collaboration networks, he had already written several papers on small worlds that are now considered classics. As our computer was churning out the first results, he posted on the Internet his first paper on collaborations between scientists. Newman's paper proved that the day-to-day business of science is conducted in densely linked clusters of scientists connected by occasional weak ties. His work, combined with our own, offered quantitative evidence for something we had felt to be true all along but that was notoriously difficult to measure before computers: Clustering is indeed present in social systems.

3.

Clustering in society is something we understand intuitively. Humans have an inborn desire to form cliques and clusters that offer familiarity, safety, and intimacy. However, a property of the social network is only of interest to scientists if it reveals something generic about most networks in nature. Therefore, Watts and Strogatz's most important discovery is that clustering does not stop at the boundary of social networks.

Though it is common to associate human intelligence with the complexity and size of the neural network within our brain, the unpronounceable *Caenorhabditis elegans*, which goes by its nickname, C. *elegans*, is living proof of how far one can get with a mere 302 neurons. Despite its two-to-three-week life span, this one-millimeter worm has had a shining career since Sydney Brenner, a prominent molecular biologist at the Molecular Sciences Institute in Berkeley, California, picked it in 1962 as a "guinea pig" of molecular biology. Since then it has been featured in thousands of articles and bred in hundreds of laboratories worldwide, and several Webpages are dedicated to it.

Though its genome is not that different from humans', C. *elegans* is one of the simplest multicellular organisms. Indeed, scientists have succeeded in figuring out the precise wiring of its nervous system, creating a map that details which neurons are connected to which. Studying this neural wiring diagram, Watts and Strogatz found that this tiny web is not much different from society at large: It displays a high degree of clustering—so high in fact that the neighbors of a neuron are five times more likely to be linked together than would be the case in a random network. The researchers detected the same pattern when studying the electricity network of the western United States, the nodes of which are generators and transformers linked together by power lines. This power network also displays an unusually high degree of clustering. So does the collaboration network of Hollywood actors, a network that we will discuss in detail in the next chapter.

Thanks to the high interest in clustering generated by Watts and Strogatz's unexpected discovery, the scientific community has subsequently scrutinized many networks. We now know that clustering is present on the Web; we have spotted it in the physical lines that connect computers on the Internet; economists have detected it in the network describing how companies are linked by joint ownership; ecologists see it in food webs that quantify how species feed on each other in ecosystems;

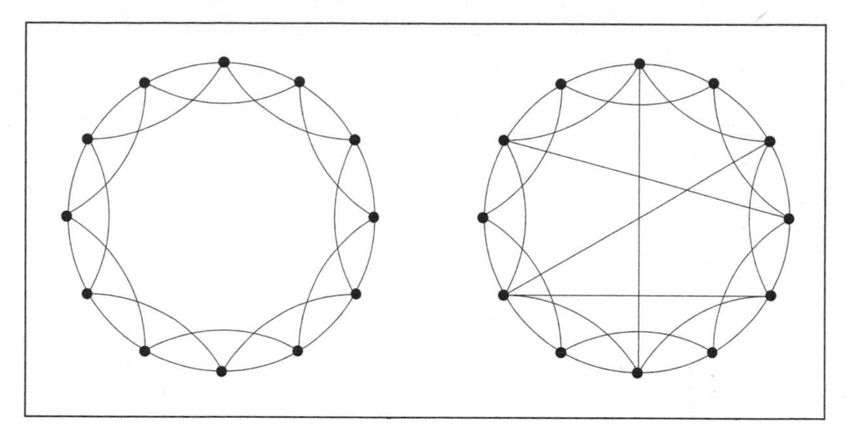

Figure 4.2 A Small and Clustered World. *To model networks with a high degree of clustering, Duncan Watts and Steven Strogatz started from a circle of nodes, where each node is connected to its immediate and next-nearest neighbors (left). To make this world a small one, a few extra links were added, connecting randomly selected nodes (right). These long-range links offer the crucial shortcuts between distant nodes, drastically shortening the average separation between all nodes.*

and cell biologists have learned that it characterizes the fragile network of molecules packed within a cell. The discovery that clustering is ubiquitous has rapidly elevated it from a unique feature of society to a generic property of complex networks and posed the first serious challenge to the view that real networks are fundamentally random.

4.

To explain the ubiquity of clustering in most real networks, Watts and Strogatz offered an alternative to Erdős and Rényi's random network model in their 1998 study published in *Nature*. They proposed a model that for the first time reconciled clustering with the completely haphazard character of random graphs. They envisioned that people live on a circle (Figure 4.2) along which everybody knows their immediate neighbors. In this simple model each node has exactly four neighbors, who are connected to *each other* by three links. Thus, the resulting network has a high clustering coefficient. Indeed, if all four neighbors were connected to each other, there would be six links between them. Since there are only three, the clustering coefficient is 3/6, or 0.5, close to the 0.56 we found for mathematicians. To see that this indeed represents significant clustering, consider a random network in which a typical node still has four neighbors but is connected randomly to any node in the system. The number of links between my four neighbors now depends on the size of the network. If I have twelve nodes, as I do in the figure, the clustering coefficient is 0.33. For 1 billion nodes, however, it drops to four over a billion! Clearly the clustering coefficient of 0.5, which the new model predicts, is gigantic compared to these values.

We have to pay a price, however, for the high clustering the model offers us. Our small world is gone. In the model society shown in the figure only my immediate and next nearest neighbors are close to me. To get to somebody on the other side of the circle, I would literally have to go around the circle, shaking innumerable hands along the way. Indeed, it is easy to check that the shortest path connecting the top node to the bottom one is at least three links long. That may not sound like a lot, but if I had had the patience (and space) to draw 6 billion nodes along the same ring, each connected to its immediate and next neighbors, the shortest path to the opposite side of the circle would require more than a billion handshakes! Thus a society on a circle not only is highly clustered but represents a very large world, as well.

In reality we all have links to distant people around the globe. Each of us has friends who do not live next door to us. If I want to find a path to a person in Australia, I will not go door to door, since sooner or later I will hit the Pacific Ocean. Instead, I remember that my best friend from high school moved to Sydney a few years ago. Thus all I need to do is find a link to my Australian target through the increasingly dense friendship links my friend is creating around him right now. A realistic model of society today must allow for distant links. We can easily achieve this in the model described above by adding a few links to some randomly chosen nodes around the circle. That is, pick any two nodes along the circle and connect them with a new link. This will decrease to one the distance between the selected nodes, and their immediate neighbors will be a lot closer to each other too. If I add many such random links, I can bring all the nodes very close together.

The surprising finding of Watts and Strogatz is that *even a few* extra links are sufficient to drastically decrease the average separation between the nodes. These few links will not significantly change the clustering coefficient. Yet thanks to the long bridges they form, often connecting nodes on the opposite side of the circle, the separation between all nodes spectacularly collapses. The model's ability to severely decrease the separation while keeping the clustering coefficient practically unchanged indicates that we can afford to be very provincial in choosing our friends, as long as a small fraction of the population has some long-range links. According to the insight provided by this simple model, six degrees are rooted in the fact that a few people have friends and relatives that do not live next door any longer. These distant links offer us short paths to people in very remote areas of the world. Huge networks do not need to be full of random links to display small world features. A few such links will do the job.

5.

The publication of the Watts-Strogatz paper on clustering, two years after Erdős's death, garnered enormous interest among physicists and mathematicians alike. First, it formalized Granovetter's vision by offering a model that did display significant clustering. Second, it played a unique role in bringing the small-world problem, a much investigated issue within sociology, to the attention of the physics and mathematics community. For a short time it seemed as if the more general, cluster-friendly model of Watts and Strogatz would replace the random universe of Erdős and Rényi. We could all relate to the simple picture of familiar local order sprinkled with a few distant links, offering a lucid explanation of the small worlds around us. The model offered an elegant compromise between the completely random world of Erdős and Rényi, which is a small world but hostile to circles of friends, and a regular lattice, which displays high clustering but in which nodes are far from each other.

Today we understand that the Watts-Strogatz model is not incompatible with the Erdős-Rényi worldview. To be sure, by assuming that we start with a regular lattice, it does allow for clusters. But in many ways, its fundamental philosophy continues to follow closely the Erdős-Rényi vision. Indeed, apart from the initial arranging of the nodes along a circle, we connect the nodes completely *randomly* to each other. Therefore both models depict a deeply egalitarian society, whose links are ruled by the throw of a dice.

When the landmark paper of Watts and Strogatz was published in 1998, my research group was trying to understand the structure of complex networks, focusing mainly on the Word Wide Web. It took us a while to fully grasp the important message of the paper and to appreciate the new model's ability to bring together the Erdős-Rényi worldview with Granovetter's clustered society. By the time it finally sank in, we had an emergency at hand. Our tiny robot returned from the Web with a network that was drastically different from the predictions of both the Erdős-Rényi and the Watts-Strogatz models. As we will see in the next chapter, it carried home a bunch of hubs—nodes with an extraordinarily large number of links. The problem was that in the egalitarian model of Erdős and Rényi such hubs are extremely rare; thus it was clear that the model could not account for our robot's finding. The Watts-Strogatz model did not fare much better: It, too, forbids nodes with significantly more links than the average node has. Something important was clearly missing from *both* models, limiting our understanding of the weblike universe. The data prompted us to search for a better understanding of real networks, eventually forcing us to abandon altogether the random worldview. Following this path, the events took a very unexpected turn. We had to give up just about everything we had learned about networks thus far.

THE FIFTH LINK

Hubs and Connectors

MALCOLM GLADWELL, A STAFF WRITER at the *New Yorker* magazine, describes in his recent book, *The Tipping Point*, a simple test to measure how social you are. He gives you a list of 248 surnames compiled from a Manhattan phone book and asks you to give yourself a point if you know anybody with that name. Multiples count, too: If you know three people named Jones, one of the names on the list, you get 3 points. Running the list by college students in the City College of Manhattan, most of them recent immigrants in their early twenties, Gladwell recorded an average score of 21. In other words, they typically knew about twenty-one people with the same surname as somebody on the list. A random group of mostly white, highly educated academics scored around 39, almost double that of the college students. This was hardly surprising. But what caught Gladwell's attention was the range. In the college class, the scores ranged from 2 to 95. In a random sample, the low score was 9 and the high score was 118. Even for a highly homogenous group of people of similar age, education, and income the range was enormous: The lowest score was 16 while the highest was 108. Gladwell ended up testing about four hundred people altogether, finding a few high scorers in every social group he looked at. His conclusion was unavoidable: "Sprinkled among every walk of life . . . are a handful of people with a truly extraordinary knack of making friends and acquaintances. They are connectors."

Connectors are an extremely important component of our social network. They create trends and fashions, make important deals, spread fads, or help launch a restaurant. They are the thread of society, smoothly bringing together different races, levels of education, and pedigrees. In noticing connectors, Gladwell thought that he was seeing something particularly human. In fact, unknown to him, he had stumbled across something altogether bigger, a phenomenon that was puzzling my research group well before the publication of *The Tipping Point*. Connectors—nodes with an anomalously large number of links—are present in very diverse complex systems, ranging from the economy to the cell. They are a fundamental property of most networks, a fact that intrigues scientists from disciplines as disparate as biology, computer science, and ecology. Their discovery has turned everything we thought we knew about networks on its head. Clustering exposed the first crack in the Erdős-Rényi random worldview. The simple model of Watts and Strogatz, discussed in the previous chapter, saved the day, reconciling the circle of friends with six degrees of separation. The connectors are the final blow to both models. Accounting for these highly connected nodes requires abandoning once and for all the random worldview.

1.

Cyberspace embodies the ultimate freedom of speech. Some may be offended, others may love it, but the content of a Webpage is hard to censor. Once posted, it is available to hundreds of millions of people. This unparalleled license of expression, coupled with diminished publishing costs, makes the Web the ultimate forum of democracy; everybody's voice can be heard with equal opportunity. Or so insist constitutional lawyers and glossy business magazines. If the Web were a random network, they would be right. But it is not. The most intriguing result of our Web-mapping project was the *complete* absence of democracy, fairness, and egalitarian values on the Web. We learned that the topology of the Web prevents us from seeing anything but a mere handful of the billion documents out there.

When it comes to the Web, the key question is no longer whether your views can be published. They can. Once published, they will be instantaneously available to anyone around the world with an Internet connection. Rather, faced with a jungle of a billion documents, the question is, if you post information on the Web, will anybody notice it?

In order to be read you have to be visible, a truism equally valid for fiction writers and scientists. On the Web the measure of visibility is the number of links. The more *incoming links* pointing to your Webpage, the more visible it is. If each document on the Web had a link to your Webpage, in a very short time everyone would know what you had to say. But the average Webpage only has about five to seven links, each pointing to one of the billion pages out there. Therefore, the likelihood that a typical document links to your Webpage is close to zero.

This conclusion applies perfectly to my homepage, www.nd.edu/~alb. According to AltaVista, there are about forty other pages worldwide linking to it. Frankly, that is a lot, considering its narrow scope. But if you take into account that there are about a billion pages to choose from, the likelihood of your discovering my Webpage is roughly forty in a billion. That is, if you randomly surf the Web and a visit to each Webpage lasts only ten seconds, you will be surfing eight years, day and night, before you run across a link pointing to my home page.

Each of us has very different interests, values, beliefs, and tastes. The links we add to our Webpages reflect this diversity. We link all over the map, from sites on African tribal art to e-commerce portals. Considering the billion plus nodes from which we can choose, we might expect the resulting linking pattern to look fairly random. Such random linking would imply that the Erdős-Rényi model reigns. A random Web would be the ultimate carrier of egalitarianism, since the Erdős-Rényi theory guarantees that all nodes are very similar to each other, each having roughly the same number of incoming links.

Our measurements, however, defied these expectations. The map returned by our robot offered evidence of a high degree of unevenness in the Web's topology. Of the 325,000 pages on the University of Notre Dame's domain we investigated, 270,000, or 82 percent of all pages, had *three or fewer* incoming links. However, a small minority, about 42 pages, had been referenced by over a thousand other pages and had more than 1,000 incoming links! Subsequent measurements on a sample of 203 million Webpages uncovered an even wider spectrum: The vast majority, as many as 90 percent of all documents, have ten or fewer links pointing to them, while a few, about three, are referenced by close to a million other pages!

Just as in society a few connectors know an unusually large number of people, we found that the architecture of the World Wide Web is dominated by a few very highly connected nodes, or *hubs*. These hubs, such as Yahoo! or Amazon.com, are extremely visible—everywhere you go, you see another link pointing to them. In the network behind the Web many unpopular or seldom noticed nodes with only a small number of links are held together by these few highly connected Websites.

The hubs are the strongest argument against the utopian vision of an egalitarian cyberspace. Yes, we all have the right to put anything we wish on the Web. But will anybody notice? If the Web were a random network, we would all have the same chance to be seen and heard. In a collective manner, we somehow create hubs, Websites to which everyone links. They are very easy to find, no matter where you are on the Web. Compared to these hubs, the rest of the Web is invisible. For all practical purposes, pages linked by only one or two other documents do not exist. It is almost impossible to find them. Even the search engines are biased against them, ignoring them as they crawl the Web looking for the hottest new sites.

2.

Kevin Bacon's movie *The Air Up There* was airing on television the night that Craig Fass, Brian Turtle, and Mike Ginelly, students from Albright College in Reading, Pennsylvania, had a revelation. It suddenly occurred to them that Bacon had played in so many different movies that you could connect him to just about any actor in Hollywood. Full of excitement, in January of 1994 they mailed a letter to the *Jon Stewart Show*, an irreverent celebrity talk show popular with college audiences. "We are three men on a mission. Our mission is to prove to the Jon Stewart audience, nay, the world, that Bacon is God." Much to their surprise, they got their fifteen minutes of fame. They were invited to appear on the Stewart show with Kevin Bacon, and charmed the audience with their ability to connect Bacon to any actor whose name was thrown at them. Down the line, however, they were desperately wrong. Bacon is no closer to the center of Hollywood than to the center of the universe.

The genius, if it can be called that, of these three students was their observation that every actor in Hollywood could be connected to Kevin Bacon with typically two to three links. For example, Tom Cruise is only one step away from Bacon because they played together in *A Few Good Men*. Using the Erdős number analogy, Tom Cruise has Bacon number one. Mike Myers has two, being connected to Robert Wagner through *The Spy Who Shagged Me*. Wagner has Bacon number one thanks to *Wild Things*. But even such historical figures as Charlie Chaplin have a path to Bacon: Chaplin played with Barry Norton in *Monsieur Verdoux*, who played with Robert Wagner in *What Price Glory*, who, as we already know, is only one link away from Bacon. Thus Charlie Chaplin has Bacon number three. To further tangle the tale, Paul Erdős has a Bacon number four, by virtue of *N Is a Number*, a documentary about him in which he plays himself. In the cast was Gene Patterson, playing himself, who later had a small role in the movie *Box of Moonlight*, through which he gains Bacon number three. And since *N Is a Number* featured the crème de la crème of graph theory, many mathematicians have not only a small Erdős number but a small Bacon number too.

The Kevin Bacon game would have remained mere movie trivia had two computer science students not watched the Stewart show. Glen Wasson and Brett Tjaden, from the University of Virginia, immediately realized that determining the distance between any two actors was a viable computer science project, if one had access to a complete database of all actors and movies ever released. The Internet Movie Database, or IMDb.com, a cinephile powerhouse offering more information about actors and movies than one could ever need, was already in place. It took Wasson and Tjaden a few weeks of programming to set up The Oracle of Bacon Website, which became the unbeatable master of the game. If you type in the name of any two actors, in milliseconds it provides the shortest path between them, listing the chain of actors and movies through which they are connected. In no time the Website was receiving over 20,000 visits per day, eventually landing on *Time* magazine's list of top ten sites of 1997. Last time I checked, on August 26, 2001, it had hosted over 13,000 visitors that day alone.

3.

We can play the Kevin Bacon game because Hollywood forms a densely interconnected network in which the nodes are actors linked by the movies in which they have appeared. An actor has links to all other actors in the cast. Thus those who have played in several movies acquire links quickly. As each actor has an average of twenty-seven links, many more than the necessary *one* to make the network fully connected, six degrees is unavoidable: *Each* actor can be connected to any other actor through three links on average. Yet, as my research group has noticed when analyzing the actor network, averages do not apply here. As many as 41 percent of actors have fewer than ten links. These are the less known actors whose names appear on the movie screen after you have walked out of the theater. A tiny minority of actors, however, have far more than ten links. John Carradine collected 4,000 links to other actors during his prolific career, while Robert Mitchum acted with 2,905 colleagues during his decades on the silver screen. These exceptionally well connected actors are the hubs of Hollywood. Remove a few of them, and suddenly the paths from almost any actor to Bacon will drastically lengthen.

On the one hand, an educated guess would be that the actors who have played in the most movies are also the most connected, having the shortest distance to everybody else in Hollywood. This turns out to be true on average: The more movies an actor plays in, the shorter his or her average distance to his or her peers. On the other hand, the list of actors with the most movies fails to give us the most-connected actors and holds some surprises, too. Compiled by Hawoong Jeong, the top ten list, showing in parentheses the number of movies in which they played, looks like this: Mel Blanc (759), Tom Byron (679), Marc Wallice (535), Ron Jeremy (500), Peter North (491), T. T. Boy (449), Tom London (436), Randy West (425), Mike Horner (418), and Joey Silvera (410). I'd wager that for most of you these names are as unrecognizable as they were to us when we first looked at the list. Well, you might know Mel Blanc, the famed voice of popular and beloved animated cartoon characters like Bugs Bunny, Woody Woodpecker, Daffy Duck, Porky Pig, Tweety Pie, and Sylvester. And those over fifty years old would have seen Tom London, perhaps the most prolific B western movie actor, portray countless sheriffs, ranch owners, and henchmen. The rest of the actors on the most-prolific list, however, eluded us. In the end, after some research, we pinned them down. They are all porn stars.

This list is perhaps the most vivid demonstration that, when it comes to networks, size does not always matter. Despite the record number of movies porn stars make, they fail to be anywhere near the center of Hollywood. As networks are clustered, nodes that are linked only to nodes in their cluster could have a central role in that subculture or genre. Without links connecting them to the outside world, they can be quite far from nodes in other clusters. Thus it is rather difficult to connect an actor who has played only in porn movies and has links only to porn stars to the cast of a Martin Scorsese or Andrey Tarkovsky movie. They simply move in very different worlds. The truly central position in networks is reserved for those nodes that are simultaneously part of many large clusters. They are the actors who have played in very different genres during their careers. They are the Webpages that not only reference modern art, but have links to all domains of human inquiry. They are the people who regularly come into contact with people from diverse fields and social strata. They are the Erdőses of mathematics, who cannot be confined to a box, who take on with equal ease problems relevant to many subfields of science. They are the Leonardo da Vincis of networks, equally at home in arts and sciences.

Of course, Bacon is a prominent Hollywood actor. He has played in over forty-six movies, collecting links to more than 1,800 actors. His average separation from everyone else in Hollywood is 2.79—that is, most actors are within three links of him. This is the reason that some are so good at the Kevin Bacon game, easily connecting any other actor to him. But is Bacon the most connected actor? When Hawoong Jeong prepared the list of the thousand most-connected actors, the real hubs of Hollywood, it took us a while to find Bacon on it. We saw Rod Steiger in the number-one spot, with an average distance of 2.53 from everybody else. Donald Pleasence, with a separation of 2.54, was right behind him. Martin Sheen, Christopher Lee, Robert Mitchum, and Charlton Heston took up the next four spots, each with a separation less than 2.57. We read hundreds of names, browsing through dozens of pages, without a sign of Bacon. Eventually we discovered him towards the end of the list, at the 876th spot.

Why do we play the Kevin Bacon game then? Bacon's prominence is a historical fluke, rooted in the publicity offered by the Stewart show. *Every* actor is three links from most actors. Bacon is by no means special. Not only is he far from the center of the universe, he's far indeed from the center of Hollywood.

4.

A random universe does not support connectors. If society were random, then in Gladwell's modest social sample of four hundred people, with their average of around 39 social links, the most social person should have far fewer acquaintances than the 118 found. If the Web were a random network, the probability of there being a page with five hundred incoming links would be 10^{-99}—that is, practically zero, indicating that hubs are forbidden in a randomly linked Web. Yet the latest Web survey, covering less than a fifth of the full Web, found four hundred such pages and one document with over two million incoming links. The chance of finding such a node in a random network is smaller than the chance of locating a particular atom in the universe. If Hollywood forms a random network, Rod Steiger does not exist, as the probability of having such a well connected actor is about 10^{-120}, which is such a small number that it's hard to come up with a proper metaphor. These incredibly small numbers help explain our surprise when we first spotted hubs on the Web and in Hollywood during our early attempts to understand the structure of real networks. There was nothing to prepare us for them because they were forbidden by both the Erdős-Rényi and the Watts-Strogatz models. They simply were not supposed to exist.

The discovery that on the Web a few hubs grab most of the links initiated a frantic search for hubs in many areas. The results are startling: We now know that Hollywood, the Web, and society are not unique by any means. For example, hubs surface in the cell, in the network of molecules connected by chemical reactions. A few molecules, such as water or ademosine triphosphate (ATP), are the Rod Steigers of the cell, participating in a huge number of reactions. On the Internet, the network of physical lines connecting computers worldwide, a few hubs were determined to play a crucial role in guaranteeing the Internet's robustness against failures. Erdős is a major hub of mathematics, as 507 mathematicians have Erdős number one. According to an AT&T study, a few phone numbers are responsible for an extraordinarily high fraction of calls placed or received. While those with a teenager living in their homes might have suspicions about the identity of some of these phone hubs, the truth is that telemarketing firms and consumer service numbers are probably the real culprits. Hubs appear in most large complex networks that scientists have been able to study so far. They are ubiquitous, a generic building block of our complex, interconnected world.

5.

Lately hubs are enjoying exceptional attention. Celebrating the power of connectors, Emanuel Rosen spends several chapters in his book *The Anatomy of Buzz* categorizing social hubs and inspecting their role in spreading news and hype. Every four years the United States inaugurates a new social hub—the president. Indeed, Franklin Delano Roosevelt's appointment book had about 22,000 names in it, making him one of the biggest hubs of his era. Three prominent biologists have recently suggested in the prestigious scientific journal *Nature* that the hublike nature of a certain molecule, the p53 protein, is the key to understanding the processes behind many forms of cancer at the molecular level. Ecologists believe that the hubs of food webs are the keystone species of an ecosystem, paramount in maintaining the ecosystem's stability.

The attention to hubs is well deserved. Hubs are special. They dominate the structure of all networks in which they are present, making them look like small worlds. Indeed, with links to an unusually large number of nodes, hubs create short paths between any two nodes in the system. Consequently, while the average separation between two randomly selected people on Earth is six, the distance between anybody and a connector is often only one or two. Similarly, while two pages on the Web are nineteen clicks away, Yahoo.com, a giant hub, is reachable from most Webpages in two to three clicks. From the perspective of the hubs the world is indeed very tiny.

The view that networks are random, held for decades under the influence of Erdős and Rényi, has lately been questioned on many fronts. Watts and Strogatz's model offered a simple explanation of clustering, bringing random networks and clustering under the same roof. Hubs, however, again challenge the status quo. They cannot be explained by either of the models we have seen so far. Therefore, hubs force us to reconsider our knowledge of networks and to ask three fundamental questions: How do hubs appear? How many of them are expected in a given network? Why did all previous models fail to account for them?

During the last two years we have answered most of these questions. Indeed, we have found that hubs are not rare accidents of our interlinked universe. Instead, they follow strict mathematical laws whose ubiquity and reach challenge us to think very differently about networks. Uncovering and explaining these laws has been a fascinating roller coaster ride during which we have learned more about our complex, interconnected world than was known in the last hundred years.

THE SIXTH LINK

The 80/20 Rule

VILFREDO PARETO, THE INFLUENTIAL ITALIAN ECONOMIST, while giving a talk in the early 1900s at an economics conference in Geneva, was repeatedly and noisily interrupted by his powerful colleague Gustav von Schmoller. Von Schmoller, who from his throne at the University of Berlin ruled the German academic world, apparently kept shouting in a patronizing tone, "But are there laws in economics?"

Despite his aristocratic upbringing Pareto had little respect for appearances, reportedly having written his monumental work *Trattato di Sociologia Generale* while owning a single pair of shoes and one suit. It was therefore easy for him to transform himself into a beggar the next day and approach von Schmoller on the street. "Please, sir," Pareto said, "can you tell me where I can find a restaurant where you can eat for nothing?" "My dear man," replied van Schmoller, "there are no such restaurants, but there is a place around the corner where you can have a good meal very cheaply." "Ah," said Pareto, laughing triumphantly, "so there *are* laws in economics!"

Pareto had turned his attention to economics after working for two decades as a railway engineer. Deeply influenced by the mathematical beauty of Newtonian physics, he devoted the rest of his life to his dream of turning economics into an exact science, describable by laws comparable in reach and universality to those formulated in Isaac

Newton's *Principia*. The fruit of his relentless pursuit, the three-volume *Trattato*, continues to be a source of inspiration and interpretation for economists and sociologists alike.

Outside academia Pareto is best known for one of his empirical observations. An avid gardener, he noticed that 80 percent of his peas were produced by only 20 percent of the peapods. A careful observer of economic inequalities, he saw that 80 percent of Italy's land was owned by only 20 percent of the population. More recently, Pareto's Law or Principle, known also as the 80/20 rule, has been turned into the Murphy's Law of management: 80 percent of profits are produced by only 20 percent of the employees, 80 percent of customer service problems are created by only 20 percent of consumers, 80 percent of decisions are made during 20 percent of meeting time, and so on. It has morphed into a wide range of other truisms as well: For example, 80 percent of crime is committed by 20 percent of criminals.

Under different guises, the 80/20 rule describes the same phenomenon: In most cases four-fifths of our efforts are largely irrelevant. Let me contribute a few more items that approximate the 80/20 rule: 80 percent of links on the Web point to only 15 percent of Webpages, 80 percent of citations go to only 38 percent of scientists, 80 percent of links in Hollywood are connected to 30 percent of actors. Though it might be tempting to infer that the 80/20 rule applies to just about anything, that would be a gross overstatement. In reality all systems following Pareto's Law are a bit special. What sets them apart is a property that plays a key role in understanding complex networks as well.

1.

When Hawoong Jeong started building our little robot to map the Web, we had naive expectations about what the network behind the Web would look like. Guided by the insights of Erdős and Rényi, we expected to find that Webpages are connected to each other randomly. As we discussed in Chapter 2, the number of links on a Webpage should follow a peaked distribution, telling us that most documents are about equally popular. But the network our robot brought back from its journey had many nodes with a few links only, and a few hubs with an extraordinarily large number of links. The biggest surprise came when we tried to fit the histogram of the node connectivity on a so-called log-log plot. The fit told us that the distribution of links on various Webpages precisely follows a mathematical expression called a *power law*.

If you are not a physicist or mathematician, most likely you have never heard of power laws. That is because most quantities in nature follow a bell curve, a distribution rather similar to the peaked distribution characterizing random networks. For example, if you measure the height of all your adult male acquaintances and prepare a histogram counting how many of them are four, five, six, or seven feet tall, you will find that most people in your sample are between five and six feet tall. Your histogram will have a peak around these values. Indeed, unless you hang out a lot with basketball players, you will have very few seven- or eight-foot people in your sample. The same is true for shorter people: Three- or four-feet-tall individuals will be rather rare. As most quantities in nature follow such a peaked distribution, ranging from our IQs to the velocity of molecules in a gas, many people are familiar with these ubiquitous bell curves.

In the past few decades scientists have recognized that on occasion nature generates quantities that follow a power law distribution instead of a bell curve. Power laws are very different from the bell curves describing our heights. First, a power law distribution does not have a peak. Rather, a histogram following a power law is a continuously decreasing curve, implying that many small events coexist with a few large events. If the heights of an imaginary planet's inhabitants followed a power law distribution, most creatures would be really short. But nobody would be surprised to see occasionally a hundred-feet-tall monster walking down the street. In fact, among six billion inhabitants there would be at least one over 8,000 feet tall. So the distinguishing feature of a power law is not only that there are many small events but that the numerous tiny events coexist with a few very large ones. These extraordinarily large events are simply forbidden in a bell curve.[1]

Each power law is characterized by a unique *exponent*, telling us, for example, how many very popular Webpages are out there relative to the less popular ones. As in networks the power law describes the degree distribution; the exponent is often called the *degree exponent.* Our measurements indicated that the distribution of incoming links on Webpages followed a power law with a unique and well-defined degree exponent close to two. A similar power law was present when we looked at outgoing links, the degree exponent this time being slightly larger.[2]

Our tiny robot offered compelling evidence that millions of Webpage creators work together in some magic way to generate a complex Web that defies the random universe. Their collective action forces the degree distribution to evade the bell curve—a signature of random networks—and to turn the Web into a very peculiar network described by a power law. The robot failed to answer our most pressing question, however. What was it about the Web that prompted it to defy the strict predictions of random networks?

Then we realized that there was another way to approach this problem. Could it be that equally simple laws characterize most complex networks and we had not seen them because we had not looked for them before? This second line of questioning turned out to be much more fruitful. Indeed, a few months later, while analyzing the actor network behind Hollywood, we found that it too followed the

---

1. Note that there is an important qualitative difference between a power law and a bell curve when it comes to the tail of the distribution. Bell curves have an exponentially decaying tail, which is a much faster decrease than that displayed by a power law. This exponential tail is responsible for the absence of the hubs. In comparison, power laws decay far more slowly, allowing for "rare events" such as the hubs.

2. This implies that the number of Webpages with exactly k incoming links, denoted by $N(k)$, follows $N(k) \sim k^{-\gamma}$, where the parameter γ is the degree exponent.

The slope of the straight line on the log-log plot indicates that the degree exponent had a value close to 2.1. When we counted how many outgoing links were on a given World Wide Web document, we observed the same pattern: The log-log plot revealed that the number of pages with exactly k outgoing links follow $N(k) \sim k^{-\gamma}$, with $\gamma = 2.5$.

same mathematical relationship: The number of actors that had links to exactly k other actors decays following a power law. Later we learned that Erdős and his mathematician colleagues obeyed this law, too. The web within the cell joined the list as we learned that the number of molecules interacting with exactly k other molecules decays following a power law. We also discovered a paper by Sid Redner, a professor of physics at Boston University, who found that the distribution of citations in physics journals follows a power law. Viewing citations as links of a network whose nodes are publications, Redner's finding implied that the citation network is also described by a power-law degree distribution. Subsequently, in numerous large networks that we and many other scientists have had a chance to investigate, an amazingly simple and consistent pattern has emerged: The number of nodes with exactly k links follows a power law, each with a unique degree exponent that for most systems varies between two and three.

2.

The striking visual and structural differences between a random network and one described by a power-law degree distribution are best seen by comparing a U.S. roadmap with an airline routing map. On the roadmap cities are the nodes and the highways connecting them the links. This is a fairly uniform network: Each major city has at least one link to the highway system, and there are no cities served by hundreds of highways. Thus most nodes are fairly similar, with roughly the same number of links. As we saw in Chapter 2, such uniformity is an inherent property of random networks with a peaked degree distribution.

The airline routing map differs drastically from the roadmap. The nodes of this network are airports connected by direct flights between them. Inspecting the maps displayed in the glossy flight magazines placed on the back of each airplane seat, we cannot fail to notice a few hubs, such as Chicago, Dallas, Denver, Atlanta, and New York, from which flights depart to almost all other U.S. airports. The vast majority of airports are tiny, appearing as nodes with at most a few links connecting them to one or several hubs. Thus, in contrast to the highway map, where most nodes are equivalent, on the airline map a few hubs connect hundreds of small airports (Figure 6.1).

A similar unevenness characterizes networks with power-law degree distribution. Power laws mathematically formulate the fact that in most real networks the majority of nodes have only *a few* links and that these numerous tiny nodes coexist with a few big hubs, nodes with an anomalously high number of links. The few links connecting the smaller nodes to each other are not sufficient to ensure that the network is fully connected. This function is secured by the relatively rare hubs that keep real networks from falling apart.

In a random network the peak of the distribution implies that the vast majority of nodes have the same number of links and that nodes deviating from the average are extremely rare. Therefore, a random network has a characteristic *scale* in its node connectivity, embodied by the average node and fixed by the peak of the degree distribution. In contrast, the absence of a peak in a power-law degree distribution implies that in a real network there is no such thing as a characteristic node. We see a continuous hierarchy of nodes, spanning from rare hubs to the numerous tiny nodes. The largest hub is closely followed by two or three somewhat smaller hubs, followed by dozens that are even smaller, and so on, eventually arriving at the numerous small nodes.

The power law distribution thus forces us to abandon the idea of a scale, or a characteristic node. In a continuous hierarchy there is no single node which we could pick out and claim to be characteristic of all the nodes. There is no intrinsic scale in these networks. This is the reason my research group started to describe networks with power-law degree distribution as *scale-free*. With the realization that most complex networks in nature have a power-law degree distribution, the term *scale-free networks* rapidly infiltrated most disciplines faced with complex webs.

Neither the hierarchy of omnipresent hubs nor the accompanying power laws were accounted for in either of the network theories available at the time that we discovered them in 1999. If anything, they were considered merely accidental. The random network theory of Erdős and Rényi and its cluster-friendly extension by Watts and Strogatz both insisted that the number of nodes with *k* links should decrease

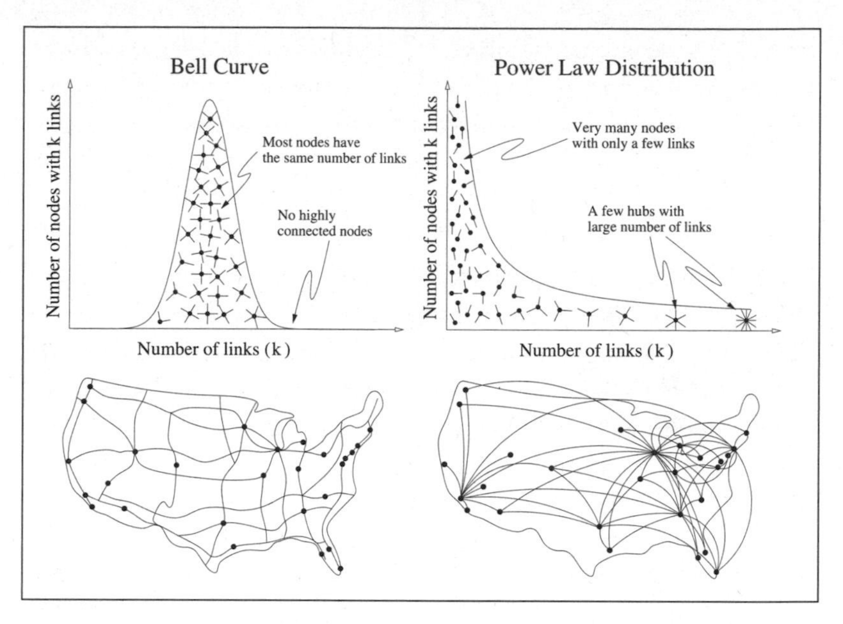

Figure 6.1 Random and Scale-Free Networks. *The degree distribution of a random network follows a bell curve, telling us that most nodes have the same number of links, and nodes with a very large number of links don't exist (top left). Thus a random network is similar to a national highway network, in which the nodes are the cities, and the links are the major highways connecting them. Indeed, most cities are served by roughly the same number of highways (bottom left). In contrast, the power law degree distribution of a scale-free network predicts that most nodes have only a few links, held together by a few highly connected hubs (top right). Visually this is very similar to the air traffic system, in which a large number of small airports are connected to each other via a few major hubs (bottom right).*

exponentially—a much faster decay than that predicted by a power law. They both told us, in rigorous mathematical terms, that *hubs do not exist.*

The surprising discovery of power laws in the Web forced us to acknowledge the hubs. The slowly decaying power law distribution accommodates such highly linked anomalies in a natural way. It predicts that each scale-free network will have several large hubs that will fundamentally define the network's topology. The finding that most networks of conceptual importance, ranging from the World Wide Web to the network within the cell, are scale-free gave legitimacy to hubs. We would come to see that they determine the structural stability, dynamic behavior, robustness, and error and attack tolerance of real networks. They stand as proof of the highly important organizing principles that govern network evolution.

3.

Pareto never used the phrase *80/20*. It was arrived at by later economists working with Pareto's observation. Rather, at the end of the nineteenth century he noticed that a few quantities in nature and the economy defy the omnipresent bell curve and instead follow a power law. His most celebrated discovery was that income distribution follows a power law, implying that most money is earned by a few very wealthy individuals, while the majority of the population earn rather small amounts. Pareto's finding implies that roughly 80 percent of money is earned by only 20 percent of the population, an inequality that is still with us a hundred years after Pareto's discovery.

It is not clear when the term *80/20* surfaced. Whereas physicists and mathematicians nonchalantly talk about power laws, the 80/20 principle pervades the popular press and business literature. But every time an 80/20 rule truly applies, you can bet that there is a power law behind it. Power laws formulate in mathematical terms the notion that a few large events carry most of the action.

Power laws rarely emerge in systems completely dominated by a roll of the dice. Physicists have learned that most often they signal a transition from disorder to order. Thus the power laws we spotted on the Web indicated, for the first time in precise mathematical terms, that real networks are far from random. Complex networks finally started to speak to us in a language that scientists trained in self-organization and complexity could finally understand. They spoke of order and emerging behavior. We just needed to listen carefully.

It might seem that the discovery that networks obey a simple power law would be exciting only to a few mathematicians or physicists. But power laws are at the heart of some of the most stunning conceptual advances in the second half of the twentieth century, emerging in fields like chaos, fractals, and phase transitions. Spotting them in networks signaled unsuspected links to other natural phenomena and placed networks at the forefront of our understanding of complex systems in general. The fact that the networks behind the Web, Hollywood, scientists, the cell, and many other complex systems all obey a power law allowed us to paraphrase Pareto and claim for the first time that perhaps *there were laws behind complex networks*.

4.

With a big O head and two great *H* ears, the Mickey Mouse–shaped water molecule and its H_2O symbol is familiar to all of us. Its size and internal structure are known in miniscule detail. This is hardly surprising: Water is the most common and most studied substance on Earth. But liquid water, the collection of billions of cohesive molecules crowded in a glass, continues to challenge us.

Gases are simple: Molecules fly in empty space, taking notice of each other only when they bounce into one another. Crystals are the opposite but relatively simple, too: Molecules hold hands tightly to create a perfectly rigid lattice. Liquids, however, strike a delicate balance between these two extremes. The attractive forces that keep the water molecules together are not strong enough to coerce them into a rigid order. Trapped between order and chaos, water molecules participate in a majestic dance in which some molecules come together, form small and somewhat ordered groups, move together, and in no time break apart to join other molecules forming yet other groups.

Chilling a glass of water does not significantly alter this magnificent water dance. It only makes the motion of molecules more dignified—heavier and slower. At 0° C, however, something special happens. The water molecules suddenly form a perfectly ordered ice crystal, like wandering soldiers lining up at an officer's command. But soldiers rehearse this drill hundreds of times, painfully learning their precise position in the formation. Water molecules, in contrast, may have never experienced ice before. They follow a mysterious urge to exchange their wandering lifestyle for a rigid, ordered one. Ice, a familiar symbol of cold and perfect order, emerges spontaneously.

Exchanging their water dance for the cold crystalline order of a solid is one of the best-known examples of a *phase transition*, a phenomenon that physicists had sought to understand for decades prior to the 1960s. Phase transitions are common in various materials, taking forms and appearances markedly different from frozen water. For example, each atom in a ferromagnetic metal has a magnetic *moment* or spin, often represented by tiny arrows piercing the atom. At high temperatures the atoms point their spins randomly in all different directions. Cooled to some critical temperature, however, all atoms orient their spins in precisely the same direction and form a magnet.

The freezing of a liquid and the emergence of a magnet are both transitions from *disorder* to *order*. Indeed, relative to the perfect order of the crystalline ice, liquid water is rather disorganized. At the freezing point it miraculously gives up this disordered state, choosing instead a state of high symmetry and order. Similarly, the randomly oriented spins in a ferromagnetic metal are in a state of chaotic disorder. They magically take up the highly ordered common orientation once cooled under a critical temperature. Such sudden transitions hold the key to a deep question about how nature works, of equal interest to scientists and philosophers alike: How does order emerge from disorder?

5.

The ordered and the disordered states of a magnet correspond to thermodynamically distinct phases of matter. Right at the transition point the system is poised to choose between these two phases, just like a climber on a crest choosing which side to go down the mountain. Undecided which way to go, the system frequently goes back and forth, and its vacillations increase near the critical point.

These vacillations have experimentally measurable consequences. Near the critical point, elements of order and disorder mix within the same material, signaling that the system explores both sides of the crest. In metals close to the transition temperature, clusters of atoms develop whose spins point in the same direction. The closer the metal gets to the critical point, the larger these ordered magnetic clusters become. The increasing amount of experimental evidence collected by physicists during the 1960s indicated that in the vicinity of the critical point several key quantities follow power laws. For example, the distance over which atoms communicate with each other, the *correlation length*, is often used as a rough measure of the cluster size. Measurements indicated that as we approach the critical point this correlation length increases following a power law characterized by a unique *critical exponent*. The closer the metal gets to the phase transition temperature, the larger the distance over which the spins know about each other. The strength of the magnet in the vicinity of the critical temperature, determined by the fraction of spins that point in the same direction, also follows a power law, with a different critical exponent.

As physicists carefully investigated how in various systems order emerges from disorder, more power laws were discovered to operate during a phase transition. The same laws emerged as liquids turned into gas once heated, or when a piece of lead turned into a superconductor once sufficiently chilled. The disorder-order transition started to display an amazing degree of mathematical consistency. The problem was that nobody knew why. Why do liquids, magnets, and superconductors lose their identity at some critical point and decide to follow identical power laws? What is behind the high degree of similarity between such disparate systems? And what do power laws have to do with it?

6.

One of the first major breakthroughs toward understanding the transition from disorder to order came during the Christmas week of 1965. Leo Kadanoff, a physicist at the University of Illinois at Urbana, had a sudden insight: In the vicinity of the critical point we need to stop viewing atoms separately. Rather, they should be considered communities that act in unison. Atoms must be replaced by boxes of atoms such that within each box all atoms behave as one.

By this time the large number of hours devoted to phase transitions, drawing the best and brightest in theoretical physics, had led to the discovery of nine different critical exponents, each associated with some power law emerging in the vicinity of the critical point. Kadanoff's idea offered an appealing visual model that could be used to derive precise mathematical relationships amongst this crowded zoo of exponents. He demonstrated that the transition from disorder to order did not require all nine unknown exponents but could be expressed in terms of any two of them. Unknown to him, several other researchers arrived at the same conclusion simultaneously. Ben Widom, a physical chemist from Cornell University, and A. Z. Patashinskii and V. L. Pokrovskii, physicists from the Soviet Union, derived similar scaling relations through a different route. A set of inequalities between the exponents, derived by physicist Michael Fisher of Cornell University, offered further hints of order within the zoo.

Something was still missing, however. There was no theory that could provide the two remaining exponents or explain why power laws appeared each time order spontaneously emerged in complex systems. It was not entirely clear if such an all-encompassing theory existed at all. Based on the beauty and self-consistency of the results obtained thus far, everyone hoped that it did. The physics community had to wait until November 1971 for the final answer. It came, unexpectedly, from a physicist with little track record in phase transitions and critical phenomena.

In the late sixties Kenneth Wilson was an assistant professor with a mixed reputation in the physics department of Cornell University. Everybody knew that he was bright, yet his brilliance had not translated into publications—the tangible measure of success in academia. This state of affairs was very close to jeopardizing his job at Cornell. Forced to publish by his tenure committee, he pulled several manuscripts from his desk drawer. Two of these, submitted simultaneously on June 2, 1971, and published in November of the same year by *Physical Review B*, turned statistical physics around. They proposed an elegant and all-encompassing theory of phase transitions.

Wilson took the scaling ideas developed by Kadanoff and molded them into a powerful theory called *renormalization*. The starting point of his approach was *scale invariance:* He assumed that in the vicinity of the critical point the laws of physics applied in an identical manner at all scales, from single atoms to boxes containing millions of identical atoms acting in unison. By giving a rigorous mathematical foundation to scale invariance, his theory spat out power laws each time he approached the critical point, the place where disorder makes room for order. Wilson's renormalization group not only called for power laws but for the first time could predict the values of the two missing critical exponents as well. With that he placed the last stone on the tip of the pyramid of phase transitions, an achievement that won him the 1982 Nobel prize in physics.

Nature normally hates power laws. In ordinary systems all quantities follow bell curves, and correlations decay rapidly, obeying exponential laws. But all that changes if the system is forced to undergo a phase transition. Then power laws emerge—nature's unmistakable sign that chaos is departing in favor of order. The theory of phase transitions told us loud and clear that the road from disorder to order is maintained by the powerful forces of self-organization and is paved by power laws. It told us that power laws are not just another way of characterizing a system's behavior. They are the patent signatures of self-organization in complex systems.

This unique and deep meaning of power laws perhaps explains our excitement when we first spotted them on the Web. It wasn't only that they were unprecedented and unexpected in the context of networks. It was that they lifted complex networks out of the jungle of randomness where Erdős and Rényi had placed them forty years earlier and dropped them into the center of the colorful and conceptually rich arena of self-organization. Gazing at the power laws that our little search engine carried home from its journey, we caught a glimpse of a new and unsuspected order within networks, one that displayed an uncommon beauty and coherence.

7.

Physicists trying to understand how magnets work and why water freezes experienced a revelation when scaling and renormalization group theory were unveiled in the late sixties and early seventies. They learned that near the critical point, just when order emerges from disorder, all quantities of interest follow power laws characterized by critical exponents. But whether we observe water going from liquid to gas, magma freezing into rock, a metal becoming a magnet, or a ceramic turning into a superconductor, the same laws always apply, generating the mysterious power laws.

We had finally learned that when giving birth to order, complex systems divest themselves of their unique features and display a universal behavior that has similar characteristics in a wide range of systems.

The ubiquity of power laws in systems undergoing a transition from disorder to order prompted my Ph.D. advisor, H. Eugene Stanley, who at Boston University leads the most active research group investigating phase transitions, to joke that in Boston there is only log-log paper. Stanley, who has been involved in all major discoveries shaping our understanding of phase transitions and universality, was referring to the plot used by scientists to detect the presence of power laws in experimental data. Indeed, wherever physicists, biologists, ecologists, materials scientists, mathematicians, or economists looked in the eighties and nineties, if self-organization reigned, power laws and universality greeted them. It appears that networks are no different: Behind the hubs there is a rather strict mathematical expression, a power law.

This brings us to the next puzzle. If power laws are the signature of systems in transition from chaos to order, what kind of transition is taking place in complex networks? If power laws appear in the vicinity of a critical point, what tunes real networks to their own critical point, allowing them to display a scale-free behavior? We had come to understand critical phenomena after physicists uncovered the mechanisms governing phase transitions; rigorous theories now allow us to calculate with high precision all quantities characterizing systems giving birth to order. But so far, in networks we had only *observed* the hubs. We now knew that they were the consequence of power laws—a hint of self-organization and order. To be sure, this was an important breakthrough, allowing us to remove networks from the realm of the random. But the most important questions, pertaining to the *mechanisms* that are responsible for the hubs and the power laws, were still unanswered. Are real networks in a continuous state of transition from disorder to order? Why do hubs appear in networks of all kinds, ranging from actors to the Web? Why are they described by power laws? Are there fundamental laws forcing different networks to take up the same universal form and shape? How does nature spin its webs?

THE SEVENTH LINK

Rich Get Richer

ONCE A PROMINENT MERCHANT PORT of the Portuguese empire, Porto today gives the impression of a forgotten city. Built where the slow-moving Duoro River wends its way to the Atlantic through the steep hills guarding the seashore, it carries the signature of a busy medieval town strategically located on an easily defensible narrow key. With its magnificent castles overlooking the river and a rich history of wine making, one might expect it to be one of the most visited cities in the world. But hidden as it is in the northwest corner of the Iberian Peninsula, few tourists make the detour. There are apparently too few fans of the distinctive full-bodied Porto vintage to awaken this great medieval city from its dreamlike state.

I visited Porto in the summer of 1999, shortly after my students and I finished our manuscript on the role of power laws on the Web. I was attending a workshop on nonequilibrium and dynamical systems organized by two professors of physics at the University of Porto, José Mendes and Maria Santos. During the summer of 1999 very few people were thinking about networks, and there were no talks on the subject during this workshop. But networks were very much on my mind. I could not help carrying with me on the trip our unresolved questions: Why hubs? Why power laws?

At that time the Web was the only network mathematically proven to have hubs. Struggling to understand it, we were searching for its distinguishing features. At the same time, we wanted to learn more about the structure of other real networks. Therefore, just before leaving for Porto, I had contacted Duncan Watts, who kindly provided us the data describing the power grid of the western United States and the C. *elegans* topology. Brett Tjaden, the former graduate student behind The Oracle of Bacon Website, now assistant professor of computer science at Ohio University in Athens, Ohio, sent us the Hollywood actor database. Jay Brockman, a computer science professor at Notre Dame, gave us data on a man-made network, the wiring diagram of a computer chip manufactured by IBM. Before I left for Europe, my graduate student Réka Albert and I agreed that she would analyze these networks. On June 14, a week after my departure, I received a long e-mail from her detailing some ongoing activities. At the end of the message there was a sentence added like an afterthought: "I looked at the degree distribution too, and in almost all systems (IBM, actors, power grid), the tail of the distribution follows a power law."

Réka's e-mail suddenly made it clear that the Web was by no means special. I found myself sitting in the conference hall paying no attention to the talks, thinking about the implications of this finding. If two networks as different as the Web and the Hollywood acting community both display power-law degree distribution, then some universal law or mechanism must be responsible. If such a law existed, it could potentially apply to all networks.

During the first break between talks I decided to withdraw to the quiet of the seminary where we were being housed. I did not get far, however. During the fifteen-minute walk back to my room a potential explanation occurred to me, one so simple and straightforward that I doubted it could be right. I immediately returned to the university to fax Réka, asking her to verify the idea using the computer. A few hours later she e-mailed me the answer. To my great astonishment, the idea worked. A simple, rich-get-richer phenomenon, potentially present in most networks, could explain the power laws we spotted on the Web and in Hollywood.

After Porto I returned briefly to Notre Dame before taking off for another month-long trip. It was clear, however, that we could not wait another month to submit our results. We had seven days to write a paper.

The eight-hour flight from Lisbon to New York seemed an ideal opportunity to prepare the first draft. As soon as the plane took off, I pulled out a laptop newly purchased before the Porto trip and frantically started typing. I was just about finished with the introduction when the flight attendant, handing a Coke to the passenger next to me, suddenly poured the entire contents of the glass onto my keyboard. Random letters flickered on the screen of my now useless laptop. But I did finish the paper on the plane, writing it out from beginning to end in longhand. A week later it was submitted to the prestigious journal *Science* only to be rejected after ten days without having undergone the usual peer review process because the editors believed that the paper did not meet the journal's standards of novelty and wide interest. By then I was in Transylvania, visiting my family and friends in the heart of the Carpathian Mountains. Disappointed but convinced that the paper was important, I did something that I had never done before: I called the editor who rejected the paper in a desperate attempt to change his mind. To my great surprise, I succeeded.

1.

The random model of Erdős and Rényi rests on two simple and often disregarded assumptions. First, we start with an inventory of nodes. Having all the nodes available from the beginning, we assume that the number of nodes is fixed and remains unchanged throughout the network's life. Second, all nodes are equivalent. Unable to distinguish between the nodes, we link them randomly to each other. These assumptions were unquestioned in over forty years of network research. But the discovery of hubs—and the power laws that describe them—forced us to abandon both assumptions. The manuscript submitted to *Science* was the first step along this path.

2.

There is one thing about the Web that everybody agrees on: It is growing. Each day new documents are added by individuals detailing their latest hobby or interest; by corporations expanding their online products and services; by governments increasingly reliant on the Web to disseminate information to citizens; by college professors publishing their lecture notes; by nonprofit organizations trying to reach those who could benefit from their services; and by thousands of dot.com companies designing flashy pages to compete for your wallet. It is estimated that within ten years the Web will host about an exabyte (10^{18}) of information spread across the planet in numerous formats, most of which are presently unknown. While the rate of this explosion will likely taper as the majority of information collected by humanity lands online, so far there are no signs of a slowdown.

With over a billion documents available today, it is hard to believe the Web emerged one node at a time. But it did. Barely a decade ago it had only *one* node, Tim Berners-Lee's famous first Webpage. As physicists and computer scientists started creating pages of their own, the original site gradually gained links pointing to it. This modest Web of a dozen primitive documents was the precursor to the planet-sized self-assemblage the Web is today. Despite its overwhelming dimensions and complexity, it continues to grow incrementally, node by node. This expansion is in stark contrast to the assumption of the network models described so far in this book, which assume the number of nodes in a network is constant over time.

The Hollywood network also started with a tiny core, the actors of the first silent movies back in the 1890s. According to the IMDb.com database, Hollywood had only 53 actors in 1900. With increasing demand for motion pictures, this core slowly expanded, adding a few new faces with each movie. Hollywood experienced its first boom between 1908 and 1914, when the number of actors joining the trade went from under 50 to close to 2,000 a year. A second spectacular boom starting in the 1980s turned moviemaking into the entertainment megaindustry we know today. From a tiny cluster of silent actors grew a gigantic network of over a half-million nodes, and it continues to grow at an incredible rate. In the period of only one year, 1998, as many as 13,209 names of actors appearing for the first time on the wide canvas of the movie screen were added to the IMDb.com database.

Despite their diversity most real networks share an essential feature: growth. Pick any network you can think of and the following will likely be true: Starting with a few nodes, it grew incrementally through the addition of new nodes, gradually reaching its current size. Obviously, growth forces us to rethink our modeling assumptions. Both the Erdős-Rényi and Watts-Strogatz models assumed that we have a fixed number of nodes that are wired together in some clever way. The networks generated by these models are therefore *static*, meaning that the number of nodes remains unchanged during the network's life. In contrast, our examples suggested that for real networks the static hypothesis is not appropriate. Instead, we should incorporate growth into our network models. This was the initial insight we gained while trying to explain the hubs. In so doing, we ended up dethroning the first fundamental assumption of the random universe—its static character.

3.

It is relatively easy to model a growing network. We start from a tiny core and keep adding nodes, one after the other. Let us assume that each new node has two links. Thus, if we start with two nodes, our third node will link to both of them. The fourth node has three nodes from which to choose. How do we pick which two we should link to? For the sake of simplicity, let's follow the lead of Erdős and Rényi and randomly select two of the three nodes and link the new node to them. We can continue this process indefinitely, so that each time we add a new node, we connect it to two randomly selected nodes. The network generated by this simple algorithm, called *Model A*, differs from the random network model of Erdős and Rényi only in its growing nature. This difference, however, is significant. Despite the fact that we choose the links randomly and democratically, the nodes in Model A are not equivalent to each other. We have easily identifiable winners and losers. At each moment all nodes have an equal chance to be linked to, resulting in a clear advantage for the senior nodes. Indeed, apart from some rare statistical fluctuations, the first nodes in Model A will be the richest, since these nodes have had the longest time to collect links.

The poorest node will be the last one to join the system, with two links only, because nobody has had time to link to it yet. Model A was among our first attempts to explain the power laws we observed on the Web and in Hollywood. The computer simulations quickly convinced us that we had not yet found the answer. The degree distribution, the function that distinguishes scale-free networks from random models, decayed too fast, following an exponential. While the early nodes were clear winners, the exponential form predicted that they are too small and there are too few of them. Therefore, Model A failed to account for the hubs and the connectors. It demonstrated, however, that growth *alone* cannot explain the emergence of power laws.

4.

During the 1999 Super Bowl numerous neverheardof.com companies such as OurBeginning.com, WebEx.com, and Epidemic Marketing blew $2 million per advertising spot to bring their name to millions of Americans following the duel between Denver and St. Louis. In one year alone E*Trade spent $300 million promoting itself. AltaVista, one of the most popular search engines, had an advertising budget close to $100 million. America Online, the Goliath of the online world, effectively matched that with $75 million. In 1999 over $3.2 billion was spent on online marketing, about half the amount spent during the same period on cable television advertising, a medium whose history spans over two decades.

What did these companies want to achieve? The answer is simple, if unconventional. Startups and established companies alike had been burning venture capital and hard-earned cash, millions a day, to defeat the random universe of Erdős and Rényi. They knew that we do not link randomly on the Web. They wanted to take advantage of this nonrandomness by begging us to link to them.

How do we in fact decide which Websites to link to on the World Wide Web? According to the random network models, we would randomly link to any of the nodes. A bit of reflection as to how we make our choices, however, indicates otherwise. For example, choices of Webpages with links to news outlets abound. A quick search for "news" on Google returns about 109,000,000 hits. Yahoo's manually ordered directory offers a choice of over 8,000 online newspapers. How do we pick one? The random network models tell us that we select *randomly* from the list. Frankly, I do not think that anybody ever does that. Rather, most of us are familiar with a few major news outlets. Without giving the matter much thought we link to one of them. As a longtime reader of the *New York Times*, it is a no-brainer for me to choose nytimes.com. Others might prefer CNN.com or MSNBC.com. Significantly, however, the Webpages to which we prefer to link are not ordinary nodes. They are hubs. The better known they are, the more links point to them. The more links they attract, the easier it is to find them on the Web and so the more familiar we are with them. In the end we all follow an unconscious bias, linking with a higher probability to the nodes we know, which are inevitably the more connected nodes of the Web. We prefer hubs.

The bottom line is that when deciding where to link on the Web, we follow *preferential attachment:* When choosing between two pages, one with twice as many links as the other, about twice as many people link to the more connected page. While our individual choices are highly unpredictable, as a group we follow strict patterns.

Preferential attachment rules in Hollywood as well. The producer whose job it is to make a movie profitable knows that stars sell movies. Thus casting is determined by two competing factors: the match between the actor and the role, and the actor's popularity. Both introduce the same bias into the selection process. Actors with more links have a higher chance of getting new roles. Indeed, the more movies an actor has made, the more likely it is that he or she will appear again on the casting director's radar screen. This is where aspiring actors have a huge disadvantage, a Catch-22 everybody knows both in and out of Hollywood. You need to be known to get good roles, but you need good roles in order to be known.

The World Wide Web and Hollywood force us to abandon the second important assumption inherent in random networks—their democratic character. In the Erdős-Rényi and Watts-Strogatz models there is no difference between the nodes of a network; thus all nodes are equally likely to get links. The examples just discussed suggest otherwise. In real networks linking is never random. Instead, popularity is attractive. Webpages with more links are more likely to be linked to again, highly connected actors are more often considered for new roles, highly cited papers are more likely to be cited again, connectors make more new friends. Network evolution is governed by the subtle yet unforgiving law of preferential attachment. Guided by it, we unconsciously add links at a higher rate to those nodes that are already heavily linked.

5.

Putting the pieces of the puzzle together, we find that real networks are governed by two laws: *growth* and *preferential attachment*. Each network starts from a small nucleus and expands with the addition of new nodes. Then these new nodes, when deciding where to link, prefer the nodes that have more links. These laws represent a significant departure from earlier models, which assumed a fixed number of nodes that are randomly connected to each other. But are they sufficient to explain the hubs and power laws encountered in real networks?

To answer this, in the 1999 *Science* paper we proposed a network model that incorporates both laws. The model is very simple, as growth and preferential attachment naturally lead to an algorithm defined by two straightforward rules (Figure 7.1):

A. *Growth:* For each given period of time we add a new node to the network. This step underscores the fact that networks are assembled one node at a time.

B. *Preferential attachment:* We assume that each new node connects to the existing nodes with two links. The probability that it will choose a given node is proportional to the number of links the chosen node has. That is, given the choice between two nodes, one with twice as many links as the other, it is twice as likely that the new node will connect to the more connected node.

Each time we repeat (a) and (b), we add a new node to the network. Therefore, node by node we generate a continuously expanding

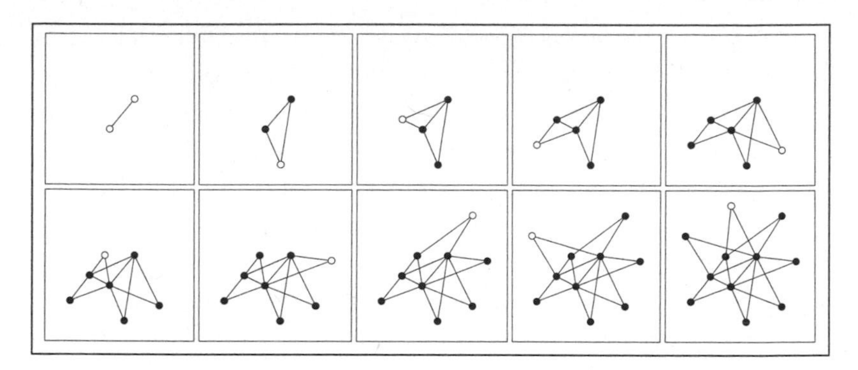

Figure 7.1 The Birth of a Scale-Free Network. *The scale-free topology is a natural consequence of the ever-expanding nature of real networks. Starting from two connected nodes (top left), in each panel a new node (shown as an empty circle) is added to the network. When deciding where to link, new nodes prefer to attach to the more connected nodes. Thanks to growth and preferential attachment, a few highly connected hubs emerge.*

web (Figure 7.1). This model, combining growth and preferential attachment, was our first successful attempt to explain the hubs. Réka's computer simulations soon indicated that it generated the elusive power laws. As the first model to explain the scale-free power laws seen in real networks, it quickly became known as the *scale-free model.*

6.

Why do hubs and power laws emerge in the scale-free model? First, growth plays an important role. The expansion of the network means that the early nodes have more time than the latecomers to acquire links: If a node is the last to arrive, no other node has the opportunity to link to it; if a node is the first in the network, all subsequent nodes have a chance to link to it. Thus growth offers a clear advantage to the senior nodes, making them the richest in links. Seniority, however, is not sufficient to explain the power laws. Hubs require the help of the second law, preferential attachment. Because new nodes prefer to link to the more connected nodes, early nodes with more links will be selected more often and will grow faster than their younger and less connected peers. As more and more nodes arrive and keep picking the more connected nodes to link to, the first nodes will inevitably break away from the pack, acquiring a very large number of links. They will turn into hubs. Thus preferential attachment induces a *rich-get-richer* phenomenon that helps the more connected nodes grab a disproportionately large number of links at the expense of the latecomers.

This rich-get-richer phenomenon naturally leads to the power laws observed in real networks. Indeed, the computer simulations we performed indicated that the number of nodes with exactly *k* links follows a power law for any value of *k*. The precise value of the degree exponent, the parameter that characterizes the power law distribution, was no longer a mystery either. We were able to calculate it analytically, using a mathematical tool, called a *continuum theory*, that we developed for this purpose. Indeed, thanks to preferential attachment, each node attracts new links at a rate proportional to the number of its current links. Using this simple observation, we were able to propose a simple equation predicting how nodes acquire links as the network expands. The solution allowed us to calculate analytically the degree distribution, confirming that indeed it follows a power law.[1]

Could either growth or preferential attachment alone explain the power laws? Computer simulations and calculations convinced us that both are necessary to generate a scale-free network. A growing network without preferential attachment has an exponential degree distribution, which is similar to a bell curve in that it forbids the hubs. In the absence of growth we are back to the static models, unable to generate the power laws.

7.

Our purpose with the scale-free model was rather modest: to demonstrate that two simple laws of *growth* and *preferential attachment* could solve the puzzle of hubs and power laws. Therefore, the model's great

---

[1] The degree exponent for the scale free model is $\gamma = 3$, i.e. the degree distribution follows $P(k) \sim k^{-3}$.

influence on subsequent research was a pleasant surprise for us, particularly since it was clear from the beginning that the topology of real networks was shaped by many effects that we had ignored for the purpose of simplicity and transparency. One of the most obvious of these is the fact that, whereas all links present in the scale-free model are added when new nodes join the network, in most networks new links can emerge spontaneously. For example, when I add to my Webpage a link pointing to nytimes.com, I create an *internal link* connecting two old nodes. In Hollywood, 94 percent of links are internal, formed when two established actors work together for the first time. Another feature absent from the scale-free model is that in many networks nodes and links can disappear. Indeed, many Webpages go out of business, taking with them thousands of links. Links can also be rewired, as when we decide to replace our link to CNN.com with a new one pointing to nytimes.com. These and other phenomena frequent in some networks but absent from the scale-free model illustrate that the evolution of real networks is far more complex than the scale-free model predicts. To understand networks in the complex world around us, we would have to incorporate these mechanisms into a consistent network theory and explain their impact on the network structure.

After submitting our paper on the scale-free model, Réka Albert and I started to investigate the effects of processes like internal links and rewiring on the structure of scale-free networks. We were no longer alone, however. A month after our paper's publication in *Science*, I learned of similar work going on in several research laboratories worldwide. Luis Amaral, my longtime collaborator, currently a research professor at Boston University, was in the process of generalizing the scale-free model to include aging, incorporating the possibility that actors stop acquiring links after retirement. Amaral, working together with Gene Stanley and two students, Antonio Scala and Mark Barthélémy, demonstrated that if nodes fail to acquire links after a certain age the size of the hubs will be limited, making large hubs less frequent than predicted by a power law. At the same time, José Mendes and Sergey Dorogovtsev were working independently on a similar problem in Porto; they soon published the first in a string of very influential papers on scale-free networks. Assuming that nodes slowly lose their ability to attract links as they age, Mendes and Dorogovtsev showed that gradual aging does not destroy the power laws, but merely alters the number of hubs by changing the degree exponent. Paul Krapivsky and Sid Redner, also from Boston University, working with Francois Leyvraz from Mexico, generalized preferential attachment to account for the possibility that linking to a node would not be simply proportional to the number of links the node has but would follow some more complicated function. They found that such effects can destroy the power law characterizing the network.

These were the first of numerous subsequent results obtained by physicists, mathematicians, computer scientists, sociologists, and biologists who scrutinized the scale-free model and its various extensions. Thanks to their efforts, we currently have a rich and consistent theory of network growth and evolution, something that would have been unthinkable just a few years ago. We understand that internal links, rewiring, removal of nodes and links, aging, nonlinear effects, and many other processes affecting network topology can be seamlessly incorporated into an amazing theoretical construct of evolving networks, which contain as a particular case the scale-free model. These processes alter the way networks grow and evolve, inevitably changing the number and the size of the hubs. But in most cases when growth and preferential attachment are simultaneously present, hubs and power laws emerge as well. In complex networks a scale-free structure is not the exception but the norm, which explains its ubiquity in most real systems.

8.

The theory of evolving networks, developed in the past three years, represents a one-way sign in network modeling. By viewing networks as dynamical systems that change continuously over time, the scale-free model embodies a new modeling philosophy. The classic static models starting with Erdős-Rényi sought simply to arrange a fixed number of nodes and links such that the final web conforms to the network being modeled. This process is similar to drawing. Seated in front of a Ferrari, our task is to draw a picture that will allow anyone to recognize the car.

Having a faithful drawing, however, doesn't bring us any closer to understanding the processes that created the car in the first place. For that we need to know how to build one just like the original. This is exactly what the various evolving network models aim to accomplish. They capture how networks are assembled by reproducing the steps followed by nature when it created its various complex systems. If we correctly model the network assembly, our final network should closely match the reality. Thus our goals have shifted from describing the topology to understanding the mechanisms that shape network evolution.

This shift in focus resulted in a dramatic change in the language of networks, as well. The static nature of the classical models had gone unnoticed until we were forced to incorporate growth. Similarly, randomness had not been a problem until the power laws required us to introduce preferential attachment. Understanding that structure and network evolution couldn't be divorced from one another made it difficult to revert to the static models that dominated our thinking for decades. These shifts in thinking created a set of opposites: *static* versus *growing*, *random* versus *scale-free*, *structure* versus *evolution*.

At the end of the previous chapter we came to an important question: Does the presence of power laws imply that real networks are the result of a phase transition from disorder to order? The answer we've arrived at is simple: Networks are not en route from a random to an ordered state. Neither are they at the edge of randomness and chaos. Rather, the scale-free topology is evidence of organizing principles acting at each stage of the network formation process. There is little mystery here, since growth and preferential attachment can explain the basic features of the networks seen in nature. No matter how large and complex a network becomes, as long as preferential attachment and growth are present it will maintain its hub-dominated scale-free topology.

The scale-free model would have remained an interesting academic exercise if there hadn't been several subsequent discoveries. The most important was the realization that most complex networks of scientific and practical importance are scale-free. The Web data was large and detailed enough to convince us that power laws can describe real networks. This realization started an avalanche of discoveries that continues to this day. As Hollywood, the metabolic network within the cell, citation networks, economic webs, and the network behind language[2] joined the list, suddenly the origins of scale-free topology became important for many scientific fields. The two laws governing network evolution built into the scale-free model offered a good starting point for exploring these diverse systems.

First, power laws gave legitimacy to the hubs. Then the scale-free model elevated the power laws seen in real networks to a mathematically backed conceptual advance. Supported by a sophisticated theory of evolving networks that allows us to precisely predict the scaling exponents and network dynamics, we have reached a new level of comprehension about our complex interconnected world, bringing us closer than ever to understanding the architecture of complexity.

But the scale-free model raised new questions. One in particular kept resurfacing: How do latecomers make it in a world in which only the rich get richer? The quest for the answer took us to a very unlikely place: the birth of quantum mechanics at the beginning of the twentieth century.

---

2. The scale-free nature of language has been shown by various research groups. In this network the nodes are words, and links represent significant cooccurences in texts, or semantic relationships (synonyms, antonyms).

THE EIGHTH LINK

Einstein's Legacy

UNLESS YOU ARE A COMPUTER SCIENTIST specializing in search engines or have paid close attention to the fortunes of dot.coms, you probably have never heard of Inktomi, the company which used to run the search engine behind the Web's most popular site, Yahoo!. Contrary to popular belief, Yahoo!, America Online, Microsoft, and many other high-profile companies don't do searches on their own. Instead, they subscribe to an enormous database such as Inktomi's, which is one of the most extensive depositories of the Web. Because Inktomi chose not to create its own portal, it never had the name recognition enjoyed by its customers and rarely made headlines. That suddenly changed during June of 2000, when the press noticed that the company's stock market value had dropped by $2.8 billion overnight. The reason? Yahoo! had fired Inktomi as its search engine, replacing it with a two-year-old startup called Google.

I met Larry Page, the Stanford dropout and cofounder of Google, in March 2000 when few people had heard of his search engine. We were speakers at a workshop sponsored by the Internet Archives in San Francisco. The event attracted an eclectic mix of computer scientists, physicists, mathematicians, librarians, lawyers, and a handful of dot.com millionaires, united in their fascination with the newly emerging online universe. Larry Page gave a short talk about his search engine and dropped a box of T-shirts in the middle of the room that proclaimed Google's signature line, "I'm Feeling Lucky." After getting home I tried mine on. I also logged onto Google and in no time became an addict. It was no surprise to me that Yahoo! wanted a part of it.

1.

Google intrigued me because it violated the basic prediction of the scale-free model, that the first mover has an advantage. In the scale-free model the most connected nodes are those that appeared first. They have had the longest time to collect links and develop into hubs. Google, launched only in 1997, was a latecomer to the Web. Popular search engines like AltaVista or Inktomi had been dominating the market long before Google's arrival, clearly making it a second mover. In less than three years, however, Google became both the biggest node and the most popular search engine.

Of course the history of business is full of stories of companies with innovative products whose consumers were hijacked by a more successful latecomer. A famous example in the computer industry is Apple, whose ingenious Newton handheld was completely obliterated by the upstart Palm. If we view products as nodes in a complex business network and consumers as links to them, we can say that Apple's links were in a short time rewired to Palm.

The aircraft industry offers a less well known example. Boeing did not invent the jet-powered passenger plane: The glory for that achievement goes to a British company, De Havilland, which started marketing the first jet, named Comet, in 1949. With a record speed of 450 miles an hour Comet captured both the European and the American markets. It did not dominate for too long, however. A year after the first commercial flights, De Havilland's planes started dropping from the sky, killing all passengers aboard. Metals wear differently on jets because of the higher speeds and altitudes. Boeing took De Havilland's tragic oversight into consideration in the design of its first jet, and five years after Comet's first flight it introduced the Boeing 707, which soon overtook De Havilland's market. Four decades later, Boeing is watching in dismay as a third mover, the European Airbus, impinges on its worldwide dominance, shrinking Boeing's market share at an unprecedented rate.

The "new kid on the block" effect appears to be present in most networks. But the scale-free model has no place for such dominant latecomers, because in the scale-free model, as in the other models we've discussed so far, all nodes are identical. To be sure, the scale-free model does differentiate between nodes based on the number of links they have acquired, which is a function of the timing of their entry into the network. But in most complex systems each node has unique characteristics that are apparent even if we do not know its connectivity. Webpages, companies, and actors have intrinsic qualities that influence the rate at which they acquire links in a competitive environment. Some show up very late and nevertheless grab all the links within a short time frame. Others rise early yet never quite make it, failing to turn their first-mover status into a hub. If we wanted to account for the fierce competition witnessed in most networks, we had to acknowledge that each node is different.

2.

Some people have a knack for turning each random encounter into a lasting social link. Some companies make a loyal partner out of every consumer. Some Webpages turn surfers into addicts. What do these nodes of society, business, and the Web have in common? Clearly, each has some intrinsic property propelling it to the head of the pack. Though it is beyond us to find a universal secret of success, we can address the *process* that separates the winners from the losers: competition in complex systems.

In a competitive environment each node has a certain *fitness*. Fitness is your ability to make friends relative to everybody else in your neighborhood; a company's competence in luring and keeping consumers compared to other companies; an actor's aptitude for being liked and remembered relative to other aspiring actors; a Webpage's ability to bring us back on a daily basis relative to the billions of other pages competing for our attention. It is a quantitative measure of a node's ability to stay in front of the competition. Fitness may have genetic roots in people; it may be related to product and management quality for companies, to talent for actors, or to content for Websites.

We can assign a fitness to each node in a network, mimicking its ability to compete for links. For example, on the Web my Webpage would have a fitness of 0.00001, while Google's would be 0.2. The actual magnitude of these numbers is irrelevant; but their ratio reflects our relative potential to lure visitors. Indeed, an average person would easily find Google about 20,000 times more useful than my personal Website.

The introduction of fitness does not eliminate growth and preferential attachment, the two basic mechanisms governing network evolution. It changes, however, what is considered attractive in a competitive environment. In the scale-free model, we assumed that a node's attractiveness was determined solely by its number of links. In a competitive environment, fitness also plays a role: Nodes with higher fitness are linked to more frequently. A simple way to incorporate fitness into the scale-free model is to assume that preferential attachment is driven by the *product* of the node's fitness and the number of links it has. Each new node decides where to link by comparing the *fitness connectivity product*[1] of all available nodes and linking with a higher probability to those that have a higher product and therefore are more attractive. Between two nodes with the same number of links, the fitter one acquires links more quickly. If two nodes have the same fitness, however, the older one still has an advantage.

This simple *fitness model*, incorporating competition and growth, was our first attempt to account for Google. It was designed as a quick fix to allow us to distinguish between nodes and give a chance to latecomers. We soon learned that fitness has far more consequences than that. Our quick fix opened an unexpected window to a rich family of phenomena that are completely invisible in an egalitarian, fitness-free universe.

3.

Ginestra Bianconi was a first-year graduate student a few months into her Ph.D. studies when I asked her to study the properties of the fitness model,

---

1. In the scale-free model the probability that a new node connects to a node with k links is given by $k/\Sigma_i k_i$. In the fitness model each node has an additional characteristic, its fitness, η. The probability to attach to a node with k links and fitness η is $k\eta/\Sigma_i k_i \eta_i$. In both expressions the sum, taken over all nodes present in the network, normalizes the probability distribution.

hoping to understand how Google turned into a hub almost overnight. Born and educated in Rome, Bianconi brought to our research group an unusual fascination for physics and a very solid background in statistical mechanics. The fitness model appeared to me to be an assignment that was relatively interesting but mathematically not terribly challenging, safe for a new student. Bianconi quickly showed me how wrong I was. First, the math behind it was far from routine. Second, it turned out to be far more interesting than anticipated. She uncovered uncountable layers of deep and surprising properties of complex networks that significantly enriched our understanding of network assembly and evolution.

Bianconi's calculations first confirmed our suspicion that in the presence of fitness the early bird is not necessarily the winner. Rather, fitness is in the driver's seat, making or breaking the hubs. In the scale-free model the connectivity of nodes in the network increases as a square root of time. The fitness model predicts a very different behavior. It tells us that nodes still acquire links following a power law, t^{β}. But the dynamic exponent, β, which measures how fast a node grabs new links, is different for each node. It is proportional to the node's fitness, such that a node that is twice as fit as any other node will acquire links faster because its dynamic exponent is twice as large. Therefore, the speed at which nodes acquire links is no longer a matter of seniority. Independent of when a node joins the network, a fit node will soon leave behind all nodes with smaller fitness. Google is the best proof of this: A latecomer with great search technology, it acquired links much faster than its competitors, eventually outshining all of them. Beauty over age.

The dynamic picture behind the scale-free model is similar to a congested one-lane highway on which each car must follow the car ahead of it. The car that enters the highway first is the inevitable winner, seniority winning over speed. The competition in the fitness model, where nodes have different fitnesses and so acquire links at different rates, is far richer. This is similar to a car race on a wide, multilane highway where cars of all different makes and models compete. The cars enter the race one after the other, each with a different engine under the hood and a driver with unique talents behind the wheel. Inevitably the race cars leave the minivans and sport utility vehicles in the dust.

The fitness model incorporating competition in complex networks posed new questions. The power law observed in the scale-free model is rooted in the fact that all nodes follow the same dynamic rule when they acquire links; in a network where some nodes grow slowly while others grab links very quickly, this careful balance is significantly disturbed. Would the power laws be present in such a competitive environment—would they apply to the fitness model? Would networks driven by competition continue to be scale-free? Or would the fierce competition for links destroy the signature of order that we had uncovered? Our search to understand how competition shapes the network's topology took us on a very unlikely detour, back in time to three giants of quantum theory: Bose, Einstein, and Planck.

4.

In June 1924 Albert Einstein received a letter and a brief manuscript, written in English, from an unknown Indian physicist from Dacca named Satyendranath Bose. Unknown to Einstein, the manuscript had been recently rejected by the *Philosophical Magazine of the Royal Society* in London. Einstein liked the manuscript so much that he set aside his own work and translated it into German, arranging for its publication in *Zeitschrift für Physik*. He even added a praising note: "In my opinion, Bose's derivation of the Planck formula signifies an important advance. The method used also yields the quantum theory of an ideal gas as I will work out in detail elsewhere."

What could get Einstein, already a Nobel prizewinner, excited enough to start work on a new problem based on an unknown physicist's unpublished manuscript? To fully understand this we have to go back two decades further. At the turn of the nineteenth century the German physicist Max Planck wanted to solve a problem of much interest to the physics community: How do objects emit light and heat? There were two competing theories, each accounting for different parts, but not all, of the experimental data. To date, the many attempts to reconcile the two approaches had been futile. Planck, in 1900, was the first to derive an expression that perfectly fit all experiments, known today as Planck's formula. He paid a very high price, however, as he had to introduce the ad hoc assumption that light and heat are emitted in small packets, or discrete quanta, an idea that disregarded his contemporaries' view that light and electromagnetic radiation are waves and are not made of discrete particles. Einstein was among the first to take Planck's hypothesis seriously. Assuming that light is indeed made of tiny particles, called photons, he predicted the *photoelectric effect*, a discovery for which he was awarded the Nobel prize in 1922. Planck, nominated by Einstein, received a Nobel in 1919 for the quantum hypothesis.

In 1924, the quantum hypothesis of light was still troublesome: A quantum mechanical derivation of Planck's formula was nonexistent. While it is a straightforward problem for undergraduate physics majors these days, at that time all attempts to derive it were unsuccessful—until Bose offered a bold solution.

What could Bose whisper from Dacca that was unknown to such titans of physics as Einstein and Planck? In the nineteenth century, physicists believed that atoms could be distinguished and numbered individually. Think of the numbered balls bouncing in a rotating drum used to pick the winning numbers in a lottery. If you pick a ball from the drum, millions of ticket holders will know exactly which one has been selected, since the numbers are painted on them. But our ability to distinguish certain subatomic particles is an illusion borrowed from daily life, argued Bose. Light particles are truly alike, unnumbered, perfectly indistinguishable. Bose showed that once statistical mechanics and thermodynamics are modified to incorporate the fact that certain subatomic particles are truly identical, Plank's law can be easily derived.

Bose's paper was still at the publisher when Einstein appeared at the Prussian Academy to present his own results, titled *Quantum Theory of Single-Atom Gases*, in which he extended Bose's method to gas molecules. Six months later he was ready with yet another publication, the *Second Treatise*. In these papers Einstein predicted a very strange phenomenon, known today as *Bose-Einstein condensation*.

At ordinary temperatures gas atoms bounce into each other at different speeds. Some are fast, others are slow. In the language of physics, some of them have high energy, others have low. If you cool the gas all the atoms slow down. To bring them to a halt, you would have to cool the gas to absolute zero, an unattainable temperature. Einstein predicted that if a gas made of indistinguishable atoms is sufficiently chilled, a significant fraction of the particles will settle to the lowest energy. That is, atoms can be forced into their lowest energy state at a critical temperature above absolute zero. When particles reach this state, they form a new form of matter called a Bose-Einstein condensate.

Einstein's 1925 prediction was greeted with great skepticism. Even the coldest spots of intergalactic space are way too hot for Bose condensation. The impossibility of reaching the required temperature—one millionth of a degree Kelvin for most atoms—made the prediction of little physical significance and questionable validity. Although spotted briefly in various systems ranging from superfluid helium to superconductors, for seventy years Einstein's predictions remained unconfirmed. Then in 1995 a group from the National Institute of Standards in Boulder, Colorado, led by Eric A. Cornell and Carl E. Weiman, cooled rubidium atoms to low enough temperatures to form a Bose-Einstein condensate.

The magnitude of Cornell and Weiman's discovery was recognized by their being awarded the Nobel prize for physics only six years later, in 2001. Their discovery not only proved in minuscule detail Einstein's prediction; it also started a revolution in atomic physics. Today we understand that Einstein's discovery applies beyond gases. Events analogous to the condensation of particles to the lowest energy level are present in many quantum systems that have little resemblance to an actual gas. Bose-Einstein condensation became a standard element of the theoretical physicists' toolkit, helping us to understand such disparate phenomena as star formation and superconductivity. It was this toolkit Bianconi reached into when she tried to understand the behavior of the fitness model.

5.

We have no subatomic particles in the World Wide Web, and networks don't have "energy levels," at least in the physicist's sense of the term. So why talk about Bose-Einstein condensation? This was the question I asked

Ginestra Bianconi on a Sunday afternoon in 2000 when I stopped briefly at the university to pick up some papers. As I was leaving my office she rather excitedly told me that she found something that might be interesting. "No time now," I had to tell her as my four-year-old son waited in the car. "See you on Monday." Bose-Einstein condensation? Who ever heard of a condensate apart from quantum mechanics? She was supposed to be working on the fitness model, ruled by the all-familiar laws of classical physics. What would quantum mechanics have to do with the Web or social networks? This was my train of thought during my two-hour drive from Notre Dame to Chicago. But I was in for a surprise on Monday.

Using a simple mathematical transformation,[2] Bianconi substituted fitness for energy, assigning an individual energy level to each node in the fitness model. Suddenly the calculations took on an unsuspected meaning: They started to resemble those that Einstein ran across eighty years earlier when he discovered the condensate. This could have been coincidental but of no consequence. But there was indeed a precise mathematical mapping between the fitness model and a Bose gas. According to this mapping, each *node* in the network corresponds to an *energy level* in the Bose gas. The larger the node's fitness, the smaller its corresponding energy level. The *links* of the network turned into *particles* in the gas, each assigned to a given energy level. Adding a new node to the network is like adding a new energy level to the Bose gas; adding a new link to the network is the same as injecting a new Bose particle into the gas. In this mapping, complex networks are like a huge quantum gas, their links behaving like subatomic particles.

This correspondence between networks and a Bose gas was highly unexpected. After all, a Bose gas is a unique creature of quantum mechanics. It is ruled by the peculiar laws of subatomic physics, which allow for a series of counterintuitive phenomena without counterparts in the macroscopic world. These laws are very different from those ruling the networks encountered throughout this book. For example, the Internet's nodes and links are macroscopic objects, routers and cables that we

2. The transformation required us to assign an energy level ε to each node with fitness η, using the expression $\varepsilon = (-1/\beta)\log \eta$, where β is a parameter that in Bose-Einstein condensation plays the role of inverse temperature.

can touch and cut if we wish to. Nobody would seriously believe that they are ruled by quantum mechanics. Yet for decades we had treated networks as geometric objects belonging strictly to the realm of mathematics. The finding that real networks are rapidly evolving dynamical systems had catapulted the study of complex networks into the arms of physicists as well. Perhaps we are in for yet another such cultural shift. Indeed, Bianconi's mapping indicated that in terms of the laws governing their behavior, networks and a Bose gas are identical. Some feature of complex networks bridges the micro- and macroworld, with consequences as intriguing as the bridge's very existence.

The most important prediction resulting from this mapping is that some networks can undergo Bose-Einstein condensation. The consequences of this prediction can be understood without knowing anything about quantum mechanics: It is, simply, that in some networks *the winner can take all.* Just as in a Bose-Einstein condensate all particles crowd into the lowest energy level, leaving the rest of the energy levels unpopulated, in some networks the fittest node could theoretically grab *all* the links, leaving none for the rest of the nodes. The winner takes all.

6.

Every network has its own fitness distribution, which tells us how similar or different the nodes in the network are. In networks where most of the nodes have comparable fitness, the distribution follows a narrowly peaked bell curve. In other networks, the range of fitnesses is very wide such that a few nodes are much more fit than most others. Google, for example, is easily tens of thousands times more interesting to all Web surfers than any personal Webpage. Indeed, the mathematical tools developed decades earlier to describe quantum gases enabled us to see that, independent of the nature of links and nodes, a network's behavior and topology are determined by the shape of its fitness distribution. But even though each system, from the Web to Hollywood, has a unique fitness distribution, Bianconi's calculation indicated that in terms of topology all networks fall into one of only *two* possible categories. In most networks the competition does not have an easily noticeable impact on the network's topology. In some networks, however, the winner takes all the links, a clear signature of Bose-Einstein condensation.

The first category includes all networks in which, despite the fierce competition for links, the scale-free topology survives. These networks display a *fit-get-rich* behavior, meaning that the fittest node will inevitably grow to become the biggest hub. The winner's lead is never significant, however. The largest hub is closely followed by a smaller one, which acquires almost as many links as the fittest node. At any moment we have a hierarchy of nodes whose degree distribution follows a power law. In most complex networks, the power laws and the fight for links thus are not antagonistic but can coexist peacefully.

In networks belonging to the second category, the *winner takes all*, meaning that the fittest node grabs *all* links, leaving very little for the rest of the nodes. Such networks develop a *star* topology, in which all nodes are connected to a central hub. In such a hub-and-spokes network there is a huge gap between the lonely hub and everybody else in the system. Thus a winner-takes-all network is very different from the scale-free networks we encountered earlier, where there is a hierarchy of hubs whose size distribution follows a power law. A winner-takes-all network is not scale-free. Instead there is a *single hub* and many *tiny* nodes. This is a very important distinction. In fact, Google's rapid rise is not an indication of winner-takes-all behavior; it only tells us that the fit get rich. To be sure, Google is one of the fittest hubs. But it never succeeded in grabbing *all* links and turning into a star. It shares the spotlight with several nodes whose number of links is comparable to Google's. When the winner takes all, there is no room for a potential challenger.

Are there any real networks that display true winner-takes-all behavior? We can now predict whether a given network will follow the fit-get-rich or winner-takes-all behavior by looking at its fitness distribution. Fitness, however, remains a somewhat elusive quantity, since the tools to precisely measure the fitness of an individual node are still being developed. But winner-takes-all behavior has such a singular and visible impact on a network's structure that, if present, it is hard to miss.

It destroys the hierarchy of hubs characterizing the scale-free topology, turning it into a starlike network, with a single node grabbing *all* the links. And there is a network in which we cannot fail to notice one node that carries the signature of a Bose-Einstein condensate. The node is called Microsoft.

7.

The most visible product of the Bill Gates and Paul Allen partnership is unquestionably Microsoft Windows. Its impact on our computer-dominated world is almost impossible to quantify. It marked the beginning of a cultural divide: You either love Windows or you hate it. You cannot be in between. Regardless of which camp you are in, however, you are most likely using it. But despite its ubiquity, Windows is not Bill Gates's most important invention. Far and away the most enduring legacy of the Gates-Allen collaboration was their idea of selling software. This was simply unimaginable before them. While a computer is a physical entity, software is only information, a never-ending string of zeroes and ones on a disk or a CD-ROM. Weirdest of all is the operating system, which does nothing but operate other strings of zeros and ones, forming a nonessential bridge between various applications and the hardware. Therefore, initially, Microsoft's business plan was opposed by everyone. Hackers who thought information and programs should be free to all hated it. Businesspeople were appalled by the notion of marketing something so easily copied.

As everyone knows, Windows prevailed despite the fact that Microsoft was not the first mover. When the first version of Windows came out, it looked like an ugly rip-off of Apple's revolutionary operating system. Apple, however, kept a rigid monopoly on its hardware, while the PC offered a free ride to all computer makers. Therefore PCs became the dominating platform in our computer-driven world, lifting Bill Gates and his Windows along with the tide.

Think of operating systems as nodes that compete for links—that is, users. Each time a user installs Windows on her or his computer, a link is added to Microsoft. The scale-free model would predict that the oldest operating system should also be the most popular. In that case, we should all be running the primitive DOS. In the more realistic fitness model, fitter operating systems would grab consumers from less fit operating systems, independent of their age.

If a fit-get-rich behavior of scale-free networks prevails in the marketplace, there should be a hierarchy of operating systems such that the most popular is followed closely by several less popular competitors. Such hierarchy is indeed present in most industries. Take, for example, computer makers. Based on worldwide shipments in the second quarter of 2000, Compaq had a 13 percent market share, closely trailed by Dell with 11 percent. Hewlett-Packard and IBM each had 7 percent, and Fujitsu-Siemens captured 4 percent of all sales. Other manufacturers accounted for a whopping 55 percent of sales, breaking the market into ever smaller segments. As most surveys list only the first five largest computer makers, it is hard to check if the distribution of their market share indeed follows a power law. I would not be surprised, however, to find that it does. The tight hierarchy suggests that computer makers are described by the fit-get-rich phase, in which no single node dominates the market.

In the operating systems market, however, such healthy competition and hierarchy is completely absent. True, Windows is not the only operating system out there. All Apple products continue to run Mac OS. DOS, the precursor of Windows, is still installed on a large number of PCs. Linux, the free-for-all operating system and the only serious challenger to Microsoft's domination, is increasingly gaining market share. UNIX continues to run most computers devoted to number crunching, used exclusively by scientists and network engineers. But all these operating systems are dwarfed by Windows's shadow, as its different versions are humming on a whopping 86 percent of all PCs. The second most popular operating system, Mac OS by Apple, has only a 5 percent market share. The ancient DOS follows closely with 3.8 percent, followed by Linux with 2.1. All other operating systems, including UNIX, capture less than 1 percent of the market.

Essentially Microsoft takes it all. As a node, it is not just slightly bigger than its next competitor. In the number of its consumers it simply cannot be compared. We all behave like extremely social Bose particles, convenience condensing us into a faceless mass of Windows users. As we purchase new computers and install Windows, we carefully feed and maintain the condensate developed around Microsoft. The operation systems market carries the classic signatures of a network that has undergone Bose-Einstein condensation, displaying clear winner-takes-all behavior. While many operating systems compete for visibility and market share, Microsoft is locked in the position of a condensate, a star dominating the vast majority of links to consumers.

8.

Nodes always compete for connections because links represent survival in an interconnected world. In most cases this competition is overt, as when companies compete for consumers, actors strive for opportunities to perform, people vie for social links. In other systems the dynamic is subtler. Molecules in a cell, for example, gain links for the benefit of the organism as a whole. But, like it or not, we are all part of a complex competitive game. As we hail some nodes and vote others out, there will always be winners and losers. The networks around us carry the signature of this competition in the stratum of links and nodes.

As long as we thought of networks as random, we modeled them as static graphs. The scale-free model reflects our awakening to the reality that networks are dynamic systems that change constantly through the addition of new nodes and links. The fitness model allows us to describe networks as *competitive* systems in which nodes fight fiercely for links. Now Bose-Einstein condensation explains how some winners get the chance to take it all.

Do the advances obtained by acknowledging fitness toss out the scale-free model? By no means. In networks that display fit-get-rich behavior, competition leads to a scale-free topology. Most networks we have studied so far—the Web, the Internet, the cell, Hollywood, and many other real networks—belong to this category. The winner shares the spotlight with a continuous hierarchy of hubs.

Yet Bose-Einstein condensation offers the theoretical possibility that in some systems the winner can grab *all* the links. When that happens, the scale-free topology vanishes. So far among real systems, only the operations systems market, with Microsoft as its dominating hub, appears to fit the bill. Are there other systems out there displaying a similar behavior? Very likely. It will take some time, however, to recognize them all.

In just the last few years our path has led us to discover new and fascinating features of our weblike universe. By uncovering the mechanisms that govern network evolution we have grasped the universality of the arsenal of tools nature uses to create the complex world around us. Now scientists in fields ranging from cell biology to business have begun to explore the consequences of the complex topologies discovered. How do these topologies affect the stability of complex systems? How do viruses spread on real networks? How do failures cascade in emergencies? Although there is much left to learn about the structure and behavior of networks, we have started to put our recent intellectual breakthroughs to work in some truly fascinating and creative ways.

THE NINTH LINK

Achilles' Heel

THE AFTERNOON TEMPERATURE IN DENVER had soared to above 100, and hundreds of office workers were rushing from office towers to the cold breeze of their cars' air conditioners. Long lines formed at gas stations for fuel and ice, traffic lights were blank, hospitals and air traffic controllers were operating on an emergency basis only, and people trapped in elevators were pushing the alarm button in vain. "On a hot day it takes no time to turn a modern office building into an incubator," remarked an office worker. "There is no ventilation, and you can't open any windows."

It is easy to forget how dependent we are on modern technology. To appreciate it, we must witness its occasional failure, like this one in the summer of 1996, when everything powered by electricity went silent between the crest of the Rockies and the Pacific. Experts had long feared a repeat of the Great Northeast Blackout of 1965, which left 30 million people without electricity for thirteen hours. In terms of financial impact, the 1996 failure of the western power system was much more devastating. Some worry that the direction in which the power industry has been evolving might make such outages more frequent than we might expect. The California electricity crises in 2001 did little to calm these fears.

Compared to today's system, the 1965 power grid was much less connected. The state of Maine survived the northeast blackout because it operated as an island with only weak ties to the rest of powerless New England. But as the nation's electricity dependency deepened over the years, people panicked whenever the power went out. According to Alan Weisman, who wrote about this in *Harper's*, utilities learned to increase stability and decrease costs by sharing facilities and bailing one another out in emergencies. As a result, formerly islanded systems began to link up, giving rise to the biggest human-made structure on Earth, containing enough wire to reach to the moon and back.

With thousands of generators, millions of miles of lines, and over a billion loads, this huge electric animal is now so interconnected and sensitive that a single disturbance can be detected thousands of miles away. But the 1996 blackout has highlighted the underlying vulnerability of this formidable system. "Having an interconnected system really makes for more efficient use of our natural resources and keeps the cost down," said spokeswoman Lynn Baker for the Bonneville Power Administration, which oversees the power grid in the Pacific Northwest. "But it means when something goes wrong, it can cascade through the system." With over $1.5 billion in damages and lost productivity, the western blackout highlighted an often ignored property of complex networks: vulnerability due to interconnectivity.

1.

Errors and failures typically corrupt all human designs. Indeed, the failure of a single component of your car's engine could easily force you to call for a tow truck. Similarly, a tiny wiring error in your computer's circuits can mean throwing the whole computer out. Natural systems beg to be different, however. Throughout Earth's geological history, species have disappeared at a rate of one per million each year. With an estimated 3 million to 100 million species living on Earth, this means that this year somewhere between three and a hundred species will vanish. Such natural extinctions appear to cause little harm, however. Over millions of years the ecosystem has developed an amazing insensitivity to errors and failures, surviving even such drastic events as the impact of the Yucatan meteorite, which killed tens of thousands of species, including the dinosaurs. The ecosystem, therefore, displays a tolerance to errors rarely seen in human-made systems.

In general, natural systems have a unique ability to survive in a wide range of conditions. Although internal failures can affect their behavior, they often sustain their basic functions under very high error rates. This is in stark contrast to most products of human design, in which the breakdown of a single component often handicaps the whole device. Lately, scientists from all disciplines have recognized the resilience of nature's designs, raising the hope that we can exploit that convenience in human-made structures. Therefore, *robustness*—rooted in the Latin word *robus*, meaning "oak," the symbol of strength and longevity in the ancient world—is an increasingly investigated topic in many fields.

Robustness is of major concern for biologists, who want to understand how a cell survives and functions under extreme conditions and frequent internal errors. It concerns social scientists and economists addressing the stability of human organizations in the face of famine, war, and changes in social and economic policy. It is a serious issue for ecologists and environmental scientists, motivating ambitious worldwide projects to preserve the sustainability of an ecosystem threatened by the disruptive effects of industrial development. Achieving robustness is the ultimate goal for specialists in increasingly interdependent communications systems, which must maintain a high degree of readiness despite inevitable malfunctions of their components.

Most systems displaying a high degree of tolerance against failures share a common feature: Their functionality is guaranteed by a highly interconnected complex network. A cell's robustness is hidden in its intricate regulatory and metabolic network; society's resilience is rooted in the interwoven social web; the economy's stability is maintained by a delicate network of financial and regulatory organizations; an ecosystem's survivability is encoded in a carefully crafted web of species interactions. It seems that nature strives to achieve robustness through *interconnectivity*. Such universal choice of a network architecture is perhaps more than mere coincidence.

2.

In the fall of 1999 the Defense Advanced Research Projects Agency, or DARPA, circulated a call for proposals to study fault-tolerant networks. As stated in the solicitation, "the program will focus primarily on the development of new network technologies that will allow the networks of the future to be resistant to attacks and continue to provide network services." A few months after the publication of our work on the World Wide Web and scale-free networks, I was looking for funding for our research in this area. The DARPA solicitation seemed an excellent opportunity for us, since the goals of the program were along the same lines as our intended research. We were hoping that scale-free networks could play a role in understanding network robustness as well. After preparing the required proposal by the November 1 deadline, I sat down with Réka Albert and Hawoong Jeong and suggested that we not wait for DARPA's answer to start work on some of the questions we formulated in the proposal.

Node failures can easily break a network into isolated, noncommunicating fragments. For example, simultaneously closing all highways going in and out of Jacksonville and Lake City, Florida, would not only isolate these cities but make the whole Florida peninsula unreachable via highway to the rest of the United States. Such fragmentation is a well-known property of networks affected by failures, a much studied topic among mathematicians and physicists alike. In general the question is, how long will it take a network to break into pieces once we randomly remove nodes? How many routers must we remove from the Internet to break it into isolated computers that cannot communicate with each other?

Clearly, the more nodes we remove, the more likely it would be that we would isolate large chunks of nodes. Decades of research on random networks, however, had indicated that network breakdown is not a gradual process. Removing only a few nodes will have little impact on the network's integrity. Yet, if the number of removed nodes reaches a critical point, the system abruptly breaks into tiny unconnected islands. Failures in random networks offer an example of an inverse phase transition: There is a critical error threshold below which the system is relatively unharmed. Above this threshold, however, the network simply falls apart.

Motivated by the DARPA proposal, in January 2000 we performed a series of computer experiments to test the Internet's resilience to router failures. Starting from the best available Internet map, we removed randomly selected nodes from the network. Expecting a critical point, we gradually increased the number of removed nodes, waiting for the moment when the Internet would fall to pieces. To our great astonishment the network refused to break apart. We could remove as many as 80 percent of all nodes, and the remaining 20 percent still hung together, forming a tightly interlinked cluster. This finding agreed with the increasing realization that the Internet, unlike many other human-made systems, displays a high degree of robustness against router failures. Indeed, a University of Michigan–Ann Arbor study had found that at any moment hundreds of Internet routers malfunction. Despite these frequent and unavoidable breakdowns, users rarely notice significant disruptions of Internet services.

Soon it became clear that we were not witnessing a property unique to the Internet. Computer simulations we performed on networks generated by the scale-free model indicated that a significant fraction of nodes can be randomly removed from *any scale-free network* without its breaking apart. The unsuspected robustness against failures is that scale-free networks display a property not shared by random networks. As the Internet, the World Wide Web, the cell, and social networks are known to be scale-free, the results indicate that their well-known resilience to errors is an inherent property of their topology—good news for the people who depend on them.

3.

What is the source of this amazing *topological robustness?* The distinguishing feature of scale-free networks is the existence of hubs, the few highly connected nodes that keep these networks together. Failures, however, do not discriminate between nodes but affect small nodes and large hubs with the same probability. If I blindly pick ten balls from a bag in which there are 10 red and 9,990 white balls, chances are ninety-nine in a hundred that I will have only white balls in my hand. Therefore, if failures in networks affect with equal chance all nodes, small nodes are far more likely to be dismantled, since there are many more of them.

Small nodes contribute little to a network's integrity. If I close a randomly chosen airport, I will most likely be shutting down one of the numerous small sites, such as the South Bend, Indiana, airport. Its absence will hardly be noticed elsewhere in America: Without it you can still travel from New York to Los Angeles or from Santa Fe to Detroit. Only the few passengers that fly in or out of South Bend will be inconvenienced. Even if as many as ten or twenty smaller airports are simultaneously closed, only a small fraction of air travel is significantly affected.

Similarly, in scale-free networks, failures predominantly affect the numerous small nodes. Thus, these networks do not break apart under failures. The accidental removal of a single hub will not be fatal either, since the continuous hierarchy of several large hubs will maintain the network's integrity. Topological robustness is thus rooted in the structural unevenness of scale-free networks: Failures disproportionately affect small nodes.

Our computer simulations left a crucial question unanswered: Do all scale-free networks display the same degree of error tolerance? We did not have to wait long for an answer. A week before our paper's publication I received an e-mail from Shlomo Havlin, a professor of physics at Bar-Ilan University in Ramat Gan, Israel, presenting the solution. The former president of the Israeli Physical Society, Havlin is one of the world's experts on percolation theory, the field of physics that developed a set of tools that now are widely used in studies of random networks. Indeed, many of the results obtained by Erdős and Rényi have since been independently discovered by physicists studying percolation.

Havlin quickly realized that scale-free networks must have a unique response to failures. Together with his students Reuven Cohen and Keren Erez and his former student Daniel ben-Avraham, currently a physics professor at Clarkson University, they set out to calculate the fraction of nodes that must be removed from an arbitrarily chosen network, random or scale-free, to break it into pieces. On one hand, their calculation accounted for the well-known result that random networks fall apart after a critical number of nodes have been removed. On the other hand, they found that for scale-free networks the critical threshold disappears in cases where the degree exponent is smaller or equal to three. Amazingly, most networks of interest, ranging from the Internet to the cell, are scale-free and have a degree exponent smaller than three. Therefore, these networks break apart only after *all* nodes have been removed—or, for all practical purposes, never.

4.

MafiaBoy, the Montreal teen responsible for the attacks on Yahoo!, eBay, and Amazon.com, was sentenced a day after the September 11 terrorist attacks on the World Trade Center and the Pentagon. He was ordered to spend eight months in a youth detention center and to make a $250 donation to charity. Prior to the sentencing, Judge Gilled Oullet argued that "this attack weakened the entire electronic communication system." But despite this and many similar claims, MafiaBoy was at no time a threat to the Internet. While temporarily denying access to several prominent Websites, his actions never harmed the infrastructure. The consequences of his attacks could not begin to compare to the potential destruction of "Operation Eligible Receiver" two years before.

In the summer of 1997, accounts began surfacing of a National Security Agency (NSA) war game developed to test the security of the United States' electronic infrastructure. Conflicting publicity suggested that the NSA had hired anywhere from twenty-five to fifty computer specialists to execute a coordinated attack on the nation's unclassified systems, sabotaging power grids, 911 systems, and the like. Purportedly, the exercise, named "Eligible Receiver," illustrated that such concerted assaults by moderately sophisticated adversaries using readily accessible tools were plausible and potentially devastating, capable of toppling U.S. military communication systems and other critical infrastructures completely.

The fallout from MafiaBoy's actions was annoying at most, denying users Web access to the most popular online sites. Eligible Receiver's attack, on the other hand, demonstrated frightening vulnerabilities in the vital arteries of the U.S. economic and security systems. Neither attack targeted nodes at random. They aimed intuitively to decimate the hubs.

5.

Mimicking the actions of a cracker who brings down the Internet's largest hubs one after the other,[1] we embarked on a new set of experiments. Like MafiaBoy and those involved in Eligible Receiver, we no longer selected the nodes randomly but attacked the network by targeting the hubs. First, we removed the largest hub, followed by the next largest, and so on. The consequences of our attack were evident. The removal of the first hub did not break the system, because the rest of the hubs were still able to hold the network together. After the removal of several hubs, however, the effect of the disruptions was clear. Large chunks of nodes were falling off the network, becoming disconnected from the main cluster. As we pushed further, removing even more hubs, we witnessed the network's spectacular collapse. The critical point, conspicuously absent under failures, suddenly reemerged when the network was attacked. The removal of a few hubs broke the Internet into tiny, hopelessly isolated pieces.

Though the simultaneous absence of the Santa Fe and South Bend airports would be hardly noticed, the closing of Chicago's O'Hare for a few hours would be headline news, affecting the entire nation's air travel. Should some event simultaneously shut down the airports in Atlanta, Chicago, Los Angeles, and New York, even if all other airports stayed open, air travel within the United States would come to a halt within

---

1. Lately the word *cracker* is used to distinguish individuals who use their expertise to break into computer systems for malicious purposes, such as shutting them down or causing other harm. In contrast, use of the word *hacker* is positive, denoting individuals with excellent computer skills who test the limits of our online universe without harming other computers or interfering with other users.

hours. Our computer simulations indicated that we face similar problems on the Internet as well. If crackers launched a successful attack against the largest Internet hubs, the potential damage could be tremendous. This is not a consequence of bad design or flaws in Internet protocols. Such vulnerability to attack is an inherent property of all scale-free networks.

Indeed, our group observed an equally spectacular breakdown when we removed the highly connected proteins from the protein interaction network of the yeast cell. The same collapse was seen by ecologists when they deleted highly connected nodes from food webs. Two subsequent papers, one by Havlin's research group and another by Duncan Callaway from Cornell University, working together with Mark Newman, Steven Strogatz, and Duncan Watts, provided the analytical backing for this observation. They demonstrated that, when the largest nodes are removed, there is a critical point beyond which the network breaks apart. Therefore, the response of scale-free networks to attacks is similar to the behavior of random networks under failures. There is a crucial difference, however. We do not need to remove a large number of nodes to reach the critical point. Disable a few of the hubs and a scale-free network will fall to pieces in no time.

6.

A few days after we submitted our manuscript describing the error and attack tolerance of complex networks, DARPA rejected our proposal. Our paper, however, was rapidly published by *Nature* and featured on the journal's cover. While disappointed by DARPA's decision, I could not blame them. In early 2000 nobody could foresee the important role scale-free networks would play in our understanding of attack survivability and fault tolerance. At that time even the fact that the Internet was a scale-free network was known only to a few researchers, and its consequences were clearly unexplored. Only today, dozens of research projects later, are we beginning to understand the ramifications of these discoveries.

Taken together, the findings indicate that scale-free networks are not vulnerable to failures. The price of this unprecedented resilience comes in their fragility under attack. The removal of the most connected nodes rapidly disintegrates these networks, breaking them into tiny noncommunicating islands. Therefore, hidden within their structure, scale-free networks harbor an unsuspected Achilles' heel, coupling a robustness against failures with vulnerability to attack.

The coexistence of robustness and vulnerability plays a key role in understanding the behavior of most complex systems. Simulations have shown that the protein network refuses to break apart under random genetic mutations. Indeed, one can remove many nodes from this key cellular network without risk of killing the organism. If, however, a drug or an illness shuts down the genes encoding the most connected proteins, the cell will not survive. Similarly, simulations performed on food webs by Ricard V. Solé and José M. Montoya from Universitat Politecnica de Catalunya in Barcelona have shown that ecosystems can easily survive random species deletions. If, however, the highly connected keystone species are removed, the ecosystem dramatically collapses.

A much studied example is the sea otter in California. The otter all but disappeared during the nineteenth century because of excessive hunting for its pelts. After federal regulators in 1911 forbade further hunting of this lovely creature, the otter made a dramatic comeback. Because it feeds on urchins, with the increase in otters the urchin population went down. With fewer urchins around, the number of kelps, a favorite food of urchins, increased dramatically. This increased the supply of food for fish and protected the coast from erosion. Therefore, protection of only one species, a hub, drastically altered both the economy and the ecology of the coastline. Indeed, finfish dominate in coastal fisheries once dedicated to shellfish.

Although scale-free networks are vulnerable to attack, several of the largest hubs must be *simultaneously* removed to crush them. This often requires taking out as many as 5 to 15 percent of all hubs at the same time. Thus crackers might need to attack and disable several hundred Internet routers, which would be quite time-consuming. It might appear, then, that despite its Achilles' heel, the Internet's topology harbors strong defenses against both random breakdowns and malicious assaults. Unfortunately, upon close inspection it turns out that this is not really the case. As we will see next, banking on topological stability against attacks is a false insurance.

7.

Though early speculations included everything from UFOs to terrorists, the 1996 summer blackout turned out not to be the result of an organized attack. Power lines expand when heated. Heating can be caused by unusually warm weather, but power lines also easily heat up and elongate if too much power is rushed across them. On a day of record temperatures, at 15:42:37 on August 10, 1996, the Allston-Keeler line in Oregon expanded and sagged close to a tree. There was a huge flash and the 1,300-megawatt line went dead. Because electricity cannot be stored, this enormous amount of power had to be suddenly shifted to neighboring lines. The shift took place automatically, funneling the current over to lower-voltage lines, of 115 and 230 kilovolts, east of the Cascade Mountains.

These power lines were not designed, however, to carry this excess power for an extended time. Loaded up to 115 percent of their thermal ratings, they too failed. A relay broke down in the 115-kilovolt line, and the excess current overheated the overloaded Ross-Lexington line, causing it too to drop into a tree. From this moment things could only keep deteriorating. Thirteen generators at the McNary Dam malfunctioned, causing power and voltage oscillations, effectively separating the North-South Pacific Intertie near the California-Oregon border. This shattered the Western Interconnected Network into isolated pieces, creating a blackout in eleven U.S. states and two Canadian provinces.

The 1996 blackout is a typical example of what scientists often call a *cascading failure*. When a network acts as a transportation system, a local failure shifts loads or responsibilities to other nodes. If the extra load is negligible, it can be seamlessly absorbed by the rest of the system, and the failure remains effectively unnoticed. If the extra load is too much for the neighboring nodes to carry, they will either tip or again redistribute the load to their neighbors. Either way, we are faced with a cascading event, the magnitude and reach of which depend on the centrality and capacity of the nodes that have been removed in the first round.

Cascading failures are not unique to power networks. A malfunctioning router will automatically prompt Internet protocols to bypass the missing node by sending packets to other routers. If the broken router carries a large amount of traffic, its absence will place a significant burden on its neighbors. Routers do not break down under too much traffic. They simply create a queue, processing as many packets as they can and dropping the rest. Therefore, too much traffic sent to a router amounts to a denial-of-service attack. Only a small percentage of the packets will make it through. Because the sender of lost packets does not get a confirmation that its message arrived, it sends it again, escalating the congestion. Therefore, the removal of several large nodes could easily create the same catastrophic disruption on the Internet as the dropping power line in Oregon did to the power system.

Cascading failures are frequent phenomena in the economy. Indeed, many attribute the East Asian economic crisis of 1997, to be discussed in more detail in Chapter 14, to the pressure the International Monetary Fund (IMF) put on the central banks of several Pacific nations, limiting their ability to provide emergency credit to troubled banks. These banks, in turn, called their loans in from companies, turning the IMF decision, arguably the biggest financial hub, into a cascade of bank and corporation failures.

Cascading failures are a well-known phenomenon in living systems as well, affecting to an equal degree ecological habitats and the cell. Indeed, as we already saw with the sea otter, the removal of some species can initiate a chain of events that can lead to a significant reorganization of the ecosystem. Similarly, the abrupt change in the concentration of a molecule can result in a cascade of events that could result in the cell's death.

Obviously, the likelihood that a local failure will handicap the whole system is much higher if we perturb the most-connected nodes. This was supported by the findings of Duncan Watts, from Columbia University, who investigated a model designed to capture the generic features of cascading failures, such as power outages, and the opposite phenomenon, the cascading popularity of books, movies, and albums, which can also be described within the same framework. His simulations indicated that most cascades are not instantaneous: Failures can go unnoticed for a long time before starting a landslide. Attempting to decrease the frequency of such cascades has inevitable consequences, however, as those cascades that do succeed are then more disruptive.

Despite these advances, our understanding of cascading failures is rather limited. Topological robustness is a structural feature of networks. Cascading failures, however, are a dynamic property of complex systems, a relatively uncharted territory. I would not be surprised to learn that there are still undiscovered laws that govern cascading failures. The discovery of those laws could have profound implications for many fields, ranging from the Internet to marketing.

8.

The error tolerance discussed in this chapter is truly good news. Network robustness implies that, when some chemicals in our body malfunction, resulting in a rash or some other minor irritation, we will be able to carry on our normal daily functions. Network robustness explains why we rarely notice the effect of router errors and why the disappearance of a few species does not lead to an environmental catastrophe.

The price of this topological robustness, however, is extreme exposure to attacks. Taking out a hierarchy of highly connected hubs will break any system. This is bad news for the Internet, since it allows crackers to design strategies that can harm the whole infrastructure. It is bad news for our economic establishment as well, for it indicates that, by focusing on the networks behind the economy, one can design strategies to cripple it. The results of the research described in this chapter thus forced us to acknowledge that topology, robustness, and vulnerability cannot be fully separated from one another. All complex systems have their Achilles' heel. We have learned that topology matters, prompting us to better appreciate the hubs. This is the first step towards defending them.

The September 11, 2001 terrorist attacks simultaneously illustrate the power of hubs and the resilience of networks. The targets were clearly not chosen at random: They were the most visible symbols of the United States' economic power and security. Targeting them, the terrorists aimed to disrupt the hubs of global capitalism. While causing a human tragedy far greater than any other event experienced by the United States in the past two decades, the terrorists did not succeed in their biggest goal: to topple the network. They started, however, a cascade of failures that continue to ripple through the world as I write. Yet despite the decimation of the twin towers of the World Trade Center, all networks, ranging from the Internet to the tangled economic web, survived—a vivid demonstration of the fundamental differences between the vulnerability of centralized human planning and the resilience of self-organized network design.

If there is any scientific lesson to learn from the events of September 11, it is that we are still far from truly understanding the interplay between robustness and vulnerability. To be sure, scientists have recently uncovered the basic principles of robustness. We now understand the fundamental role networks play in ensuring resilience, a breakthrough that is here to stay. The crucial step, however, of turning this knowledge into practical expertise has so far eluded us. Nobody could have predicted the degree of the cascading damage that the terrorist attacks would cause. As the events unfolded, everyone watched in horror, asking the same question: What's next? How vulnerable are we? Fortunately, our understanding of failures and attacks indicates that cascading failures and local breakdowns can be addressed in the language of science. Understanding these problems thus is only a matter of focusing resources on the right questions. With the increased awareness toward issues of robustness and attacks prompted by the September 11 events, unquestionably our understanding of these issues will drastically improve.

THE TENTH LINK

Viruses and Fads

GAETAN DUGAS HAD EVERYTHING he could wish for, and he knew it. With a wardrobe culled from the trendiest shops in London and Paris, and a well-built but not muscle-bound body, he was a standout in any club. He had only to proposition in his charming French Canadian accent and he could seduce anybody he wanted. "I am the prettiest," he used to say, and his friends agreed. Lately, however, he had been avoiding the popular discos and the hottest nightclubs. His preference had shifted to the steamy mirrors and heavy air of the Bay Area bathhouses. Despite his narcissistic perfection, Dugas began developing a taste for the darker houses that revealed little of his mesmerizing physical characteristics. The long hallways of shady cubicles now made him most comfortable. One night in 1982, as he prepared to exit one of those cubicles, he switched on the lights, slowly turned towards the man he had first met a few minutes before and immediately had sex with, and pointed to the purplish spots and bumps on his face. "I've got gay cancer," he said. "I'm going to die and so are you."

Dugas, once a French Canadian flight attendant, is often called Patient Zero of the AIDS epidemic. This is not because he was the first to be diagnosed with the disease but rather because at least 40 of the 248 people diagnosed with AIDS by April 1982 had either had sex with him or with someone who had. He was at the center of an emerging complex sexual network among gay men, a web anchored between the

East and West Coasts of North America, spanning San Francisco, New York, Florida, and Los Angeles.

His central role is far from coincidental. Dugas was one of the first gay men in North America to be diagnosed with Kaposi's sarcoma. By 1983 it was increasingly clear that the illness Dugas and several hundred other gay men came down with had some infectious source, and Dugas was again one of the first patients to be told that. But he continued to insist that he had skin cancer only. As cancer is not contagious, for many years he never admitted to himself that he would pose any risk to his sexual partners. Proud of his attractiveness and sexual conquests, he later confided to health care workers the intimate details of his sexual habits. He figured that he had about 250 sexual partners a year. While some estimates put the total number of his partners as high as 20,000, his decade of promiscuity in gay clubs and bathhouses clearly put him in sexual contact with at least 2,500 people.

It is not clear whether Dugas brought AIDS to North America. He traveled frequently to France, where some of the earliest cases have been discovered, but we will never know for sure if he was infected there or in the United States. What we do know, however, is that many of the earliest cases in North America were linked to him, placing him at the root of an epidemic that by now has killed almost 20 million people.

Dugas played an important role in turning the AIDS epidemic in a few short years from an obscure and rare "gay cancer" to a North American health care crisis. He is a terrifying example of the failure of classical epidemic models and evidence of the power of hubs in our highly mobile and connected society. Indeed, when it comes to viruses and epidemics, hubs make a deadly difference.

1.

Like millions of other Americans, Mike Collins saw a picture of the controversial Florida butterfly ballot on TV the night of November 8, 2000. "Jeez, how could they not follow the arrows to the dots?" was his first reaction. "Maybe I'll draw something up that will make it

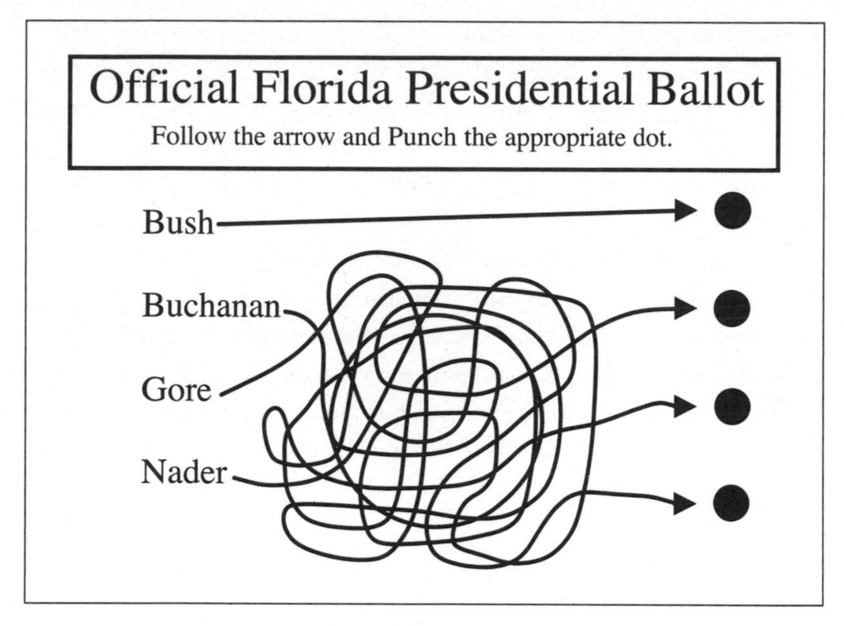

Figure 10.1 Florida's Presidential Ballot. *Mike Collins's cartoon satirizing the confusing butterfly ballots used in the 2000 presidential elections. (Reproduced with permission of Mike Collins.)*

more confusing." Collins, a twenty-six-year-old municipal water board engineer and amateur cartoonist from Elmira, New York, drew a cartoon of four lines and e-mailed it to thirty friends. The next day was his birthday and the day his sister gave birth to a daughter, so he was out all day. When he returned home that evening, a huge present awaited: 17,000 new hits on his Webpage and several hundred e-mails. While he was away his cartoon perfectly expressing everybody's frustration with the 2000 presidential election had circled the globe. Anybody who spotted it wanted a copy. Newspapers and Websites from the United States to Japan were bombarding him with requests for permission to publish. In a few hours he went from "Mike" to an instant celebrity, with girls hitting on him and parents wanting to fix him up with their daughters. While the election debate eventually died down, Collins's signature drawing became perhaps the most recognized cartoon of the decade, popping up on everything from T-shirts, which he sells through his Website, to greeting cards. It is probably the single most enduring image of the ill-fated 2000 Florida ballot.

Mike Collins's instant path to celebrity is a replay of the classic American dream. But what is unusual is the speed with which it took place. A few decades ago it was impossible to gain worldwide fame literally overnight, even in America. Something has changed. We normally credit the Internet with these changes, and certainly the medium nurtures and spreads fame. But an explanation based entirely on technology is not sufficient. We are witnessing something qualitatively new, something that is allowing ideas and fads to reach everybody with the speed of light.

2.

Gaetan Dugas and Mike Collins ostensibly have little in common. One spread a cruel disease; the other was a small-town amateur who hit it big with a clever idea. AIDS took a decade to escape its African source and permeate the world, passing from partner to partner primarily through sexual intercourse. Collins's cartoon exploded overnight, circling the world via clicks and e-mails. Nevertheless, they have something important in common. They are both examples of diffusion in a complex network. AIDS spread following the links of the intricate sexual network of the 1980s, aided by the emergence of a highly sexually active gay culture. The ballot cartoon spread instantly through the entangled network of computers, aided by our ability to reach our friends through e-mail. Both, however, followed the same fundamental laws governing the spread of fads, ideas, and epidemics in complex networks. These laws have been intensively researched by marketing executives trying to figure out how to get their product in your pocket; by sociologists seeking to understand fads, fashions, and riots; by political scientists tracking voting patterns and political fortunes; by doctors and epidemiologists hoping to curb everything from the Ebola virus to the recurring early winter flu; by teenagers writing computer viruses aiming to destroy all of Microsoft's products overnight; and by system managers determined to prevent viruses from doing just that. These laws were believed to be universal, as indeed they are. But our emerging knowledge about complex networks prompts us to see them from a new perspective.

3.

Whereas in 1933 hybrid corn was cultivated on only 40,000 acres across North America, by 1939 it had reached 24 million acres, one fourth of the nation's corn acreage. It revolutionized and reshaped American farming, eventually sweeping the whole of Midwestern agriculture in less than ten years. Iowa was particularly quick to adopt it. Though the new seed was not available before 1929, by 1939 as much as 75 percent of Iowa's corn acreage was devoted to hybrid. This rapid expansion, combined with the farmers' good bookkeeping, offered the first opportunity for researching how innovations spread. Bryce Ryan and Neal C. Cross from Iowa State College embarked on this study in 1943.

Before adopting any innovation, we normally ask ourselves several simple questions: Should I spend time evaluating the new product? Should I spend money on it? How do I know that it will work for me as promised? The questions were no different for the hybrid. To adopt it, farmers had to invest in the new seeds to replace those they already had. Though the switch promised a larger, heartier yield, there was little guarantee that the extra benefits would offset the initial investment. The risk was particularly relevant for the first adopters. Nevertheless, the hybrid took root in Iowa thanks to a small group of people willing to take risks. Today we call such people innovators.

All of us know some innovators. They are our acquaintances who jumped to buy the Apple Newton handheld computer, only to discover that the technology did not live up to its promises. A few years later they were the first to scribble characters onto the gray screen of the first Palm Pilots, this time jump-starting the handheld revolution. They are the teens who pick up on new trends before they become mainstream, the artists and intellectuals who nurture ideas well before they reach the rest of us through books, movies, and magazines. In Iowa they were the farmers for whom talking to the sales reps and reading the documentation was enough to persuade them to try the new seed.

Ryan and Cross found that plotting the number of farmers adopting the seed each year yields a curve that increases rapidly until it reaches a maximum, then drops equally fast afterwards. It is a bell curve. If a new product passes the crucial test of the innovators, based on their recommendation, the early adopters will pick it up. They are followed by the numerous early majority, until half of the people who will eventually adopt are already in the game. Beyond this point the number of new adopters starts decreasing, the innovation attracting those who are slow to make a decision but are persuaded by the overwhelming evidence in its favor. This late majority is made up of farmers who have seen half of the fields surrounding them turn over to the hybrid and are finally convinced. The curve inevitably ends with the few laggards, who join only after they have become a clear minority.

The bell curve observed by Ryan and Cross is not unique to Iowa farmers. It characterizes the spread of most innovations, offering an excellent tool by which marketing and planning experts foresee demand for a new product. However, it fails to answer something that everybody from epidemiologists to CEOs wants to know these days: What, if any, role is played by the social network in the spread of a virus or an innovation?

4.

In 1954 Elihu Katz, a researcher in the Bureau of Applied Social Research at Columbia University, circulated a proposal to study the effect of social ties on behavior. It so happened that the director of market research for the pharmaceutical giant Pfizer was a Columbia alumnus. Keen to understand how physicians adopt a new drug, he offered Katz and his two colleagues, James Coleman and Herbert Menzel, $40,000 to track the spread of tetracycline, a powerful antibiotic introduced in the mid-1950s.

Coleman, Katz, and Menzel interviewed 125 doctors from a small Illinois town, asking them to list separately the three doctors with whom they most often discussed medical practices, three from whom they sought advice regarding a medicine, and three whom they considered friends. These lists allowed them to reconstruct the complex network of social ties and influence within the medical community.

The results indicated significant differences among doctors. A few were named by a large fraction of their colleagues as playing an important role in their day-to-day decisions. They were the hubs of the medical community. The majority, however, played a much smaller role. When it came to the spread of tetracycline, the doctors named by three or more other doctors as friends were three times more likely to adopt the new drug than those who had not been named by anybody.

Using prescription records from pharmacies, the researchers could follow the spread of the drug's use along the social links. It turned out that the early adopters and early majority were predominantly doctors with numerous social links. These highly connected doctors were more likely to be in touch with innovators, thus learning about the new drug more quickly. Once adopted by these doctors, the drug spread from these hubs to their less connected colleagues, who formed the late majority. Finally came the laggards, doctors who resisted adopting the new drug until the very end.

The Pfizer study demonstrated that innovations spread from innovators to hubs. The hubs in turn send the information out along their numerous links, reaching most people within a given social or professional network. Hubs, the integral components of scale-free networks, are the statistically rare, highly connected individuals who keep social networks together. In the AIDS epidemic, the gay flight attendant Gaetan Dugas clearly qualified as a major hub. And the well-traveled Paul, with his extended circle of friends and followers, was one of the most influential hubs of early Christianity.

Hubs, often referred to in marketing as "opinion leaders," "power users," or "influencers," are individuals who communicate with more people about a certain product than does the average person. With their numerous social contacts, they are among the first to notice and use the experience of the innovators. Though not necessarily innovators themselves, their conversion is the key to launching an idea or an innovation. If the hubs resist a product, they form such an impenetrable and influential wall that the innovation can only fail. If they accept it, they influence a very large number of people.

Sociologists and marketing experts are fully aware of these opinion leaders. But until recently they treated hubs as unique phenomena, with little understanding of why and how many of them are out there. Social network models did not support the existence of hubs. The framework offered by scale-free networks has for the first time provided the legitimacy hubs deserve. As we will see, hubs are changing nearly everything we know regarding the spread of ideas, innovations, and viruses.

5.

Unveiled in 1993 as the brainchild of John Sculley, Apple's Pepsi-bred CEO, the much promoted handheld computer Newton never made it. Nevertheless, it started a revolution.

Today there are millions of pocket-sized devices in circulation. Despite this enormous number, many believe that we are still at the beginning of the bell curve for market penetration. The problem for Apple is that none of these handy devices is a Newton. Palm, Handspring, various Pocket PCs, and their countless cousins have chewed up the Apple vision, offering powerful proof that the first mover does not always have the advantage. Newton pulled together many new technologies in a "first-ever" device, promising a dream come true. It wasn't that easy, however. The nightmare started with a series of bad reviews ridiculing Newton's handwriting recognition capabilities. Critics pointed out that it voraciously consumed batteries after a mere twenty minutes of use. Disappointment followed disappointment, and sales of the MessagePad, the redesigned version of Newton, peaked at a dismal 85,000 in 1995. The product was discontinued three years later in an attempt to curtail losses after Steve Jobs was reinstalled as Apple's interim CEO.

The failure of the Newton handheld and many other products begs for an explanation. Why do some inventions, rumors, and viruses take over the globe, while others diffuse only partially or simply disappear? Why and how are losers different from winners? Clearly advertisement is not a sufficient explanation. After all, Newton failed despite Apple's enormous marketing machine. The billion-dollar question is, how does one spot the rotten apples?

6.

Aiming to explain the disappearance of some fads and viruses and the spread of others, social scientists and epidemiologists developed a very useful tool called the *threshold model*. We all differ in our willingness to accept innovation. In general, with sufficient positive evidence, each of us can be convinced to adopt a new idea. However, the level of acceptable testimony differs from one person to another. Acknowledging our differences, diffusion models assign a threshold to each individual, quantifying the likelihood that he or she will adopt a given innovation. For example, those who bought the Newton right after its release had close to a zero threshold for handheld devices. Before swiping our credit cards, however, most of us want to see a new product working; thus most of us display a higher threshold.

Despite significant differences in purpose and detail, all diffusion models predict the same phenomenon: Each innovation has a well-defined *spreading rate*, representing the likelihood that it will be adopted by a person introduced to it. For example, the spreading rate incorporates the likelihood that after being shown a new handheld, you will be prompted to buy it. Yet knowing the spreading rate alone is not sufficient to decide the fate of an innovation. For that we must calculate the *critical threshold*, a quantity determined by the properties of the network in which the innovation spreads. If the spreading rate of the innovation is less than the critical threshold, it will die out shortly. If it is over the threshold, however, then the number of people adopting it will increase exponentially until everybody who could use it does.

Recognizing that passing a critical threshold is the prerequisite for the spread of fads and viruses was probably the most important conceptual advance in understanding spreading and diffusion. Currently the critical threshold is part of every diffusion theory. Epidemiologists work with it when they model the probability that a new infection will turn into an epidemic, as the AIDS virus did. Marketing textbooks talk about it when estimating the likelihood that a product will make it in the marketplace or to understand why some never do. Sociologists use it to explain the spread of birth control practices among women. Political science exploits it to explain the life cycle of parties and movements or to model the likelihood that peaceful demonstrations will turn into riots.

For decades, a simple but powerful paradigm dominated our treatment of diffusion problems. If we wanted to estimate the probability that an innovation would spread, we needed only to know its spreading rate and the critical threshold it faced. Nobody questioned this paradigm. Recently, however, we have learned that some viruses and innovations are oblivious to it.

7.

Launched from the Philippines, Love Bug, the most damaging computer virus ever, reached every computer-literate corner of the world in hours. On May 8, 2000, the sun rose on continent after continent to the fall of computers, thousands at a time—a global domino effect sweeping from east to west. Computer security experts had hardly begun assisting the first victims in Hong Kong when system administrators of a major German newspaper watched in horror as the virus consumed 2,000 digital photographs. Spreading to Belgium, it handicapped ATM machines, denying customers vital currency. London, waking an hour later, witnessed Parliament's shutdown. Before moving on from Europe, as many as 70 percent of Swedish, German, and Dutch computers were in ruins. The carnage spread to the United States, where it sneaked into the Capitol building's computers in Washington D.C., infected 80 percent of all federal agencies, including the defense and state departments, and shut down the Bush presidential campaign's e-mail communications.

Love Bug, causing over $10 billion in damage from 45 million destroyed computers worldwide, was a well-engineered psychological booby trap that nobody could resist. How could you not immediately open a message entitled LOVE-LETTER-FOR-YOU? If you did yield to the temptation, the activated virus then erased a series of documents from your hard drive, with a particular appetite for jpeg and mp3 files that encode digital pictures and music. Next it looked for a Microsoft Outlook Express e-mail program. If it did find one, it sent new copies of the love letter to all your friends and acquaintances whose e-mail addresses you stored there.

The carnage slowed when Richard Cheng and Maricel Soriano, from the Philippines, created an antidote, a program that could immunize a computer against the bug. What is amazing about Love Bug, however, is that despite the widely and freely available antidote, the virus still exists. According to Virus Bulletin, an online resource that collects virus occurrences, Love Bug was still the seventh most active virus in April 2001, a year after the release of the program that detects and deactivates it. I received a copy as late as July 2001.

It is tempting to speculate that perhaps Love Bug is so virulent that it is virtually impossible to eradicate. But its continued presence cannot be explained by virulence alone. This was the conclusion of two physicists, Romualdo Pastor-Satorras and Alessandro Vespignani, who showed that in contrast to the solid predictions of threshold models, in real networks high virulence does not guarantee a virus's spread.

8.

The unique northern Italian town of Trieste, with its mixed and tumultuous historical heritage, is home to the prestigious International Center for Theoretical Physics. Founded and directed for decades by the Nobel prizewinning Pakistani physicist Abdus Salam, it offers a safe and intellectually challenging haven for Third World physicists, bringing them in contact with their colleagues from around the world. Romualdo Pastor-Satorras, a Spanish physicist, finished a two-year postdoctoral position at the center in 1999 before returning to Barcelona to assume a professorship. In the summer of 2000 he went back to Trieste for a two-month visit, planning to finish up several overdue projects that he and his former mentor Alessandro Vespignani had initiated during his earlier stay. While compiling the bibliography of a new manuscript, they stumbled across a computer science paper titled *Open Problems in Computer Virus Research* by Steve R. White, a computer virus expert from IBM. The paper argued that biologically inspired epidemic models do not properly describe the spread of Love Bug and other computer viruses.

Intrigued by this observation, the researchers decided to dissect the problem more carefully. Using the records of the Virus Bulletin, an online resource for computer virus prevention, they determined the likelihood that a virus would still exist several months after its first occurrence. The results were astonishing: The characteristic life of most viruses ranges between six and fourteen months. That is, viruses are infecting computers more than a year after their first occurrence and supposed eradication. As Pastor-Satorras and Vespignani put it, "these characteristic times are impressively large if compared with the interval in which antivirus software is available on the market (usually within days or weeks after the first incident report)." Like the Mummy, viruses are awakened over and over again from their sarcophagi, unable to rest.

Researchers normally use various versions of the standard threshold models to describe how computer viruses spread. In these models each computer can be either healthy or infected. During each time interval, a healthy computer can be infected by the virus if it is in contact with an already infected computer. As soon as an infected computer is cured, it becomes susceptible to infection again. Assuming that computers are connected randomly to each other, this model confirms the classical scenario of virus spread: A virulent virus, with contagiousness larger than a critical threshold, reaches most computers. In contrast, if the virus's virulence is less than the threshold, the number of newly infected computers decreases quickly, until the virus dies out.

By August 2000 Pastor-Satorras and Vespignani had concluded that White was right: Computer viruses defy the predictions of the classical epidemic models. The source of this discrepancy, however, remained unclear to them. As luck would have it, my research group's paper on the Internet's Achilles' heel was featured on the cover of *Nature* that very week. Reading it, they suddenly found the missing piece. On the Internet, computers are not connected to each other randomly. Rather, the underlying network has a scale-free topology. Thus, computer viruses should be modeled on a scale-free network instead of the random one used in all previous studies. Pastor-Satorras and Vespignani rushed to do just that, investigating for the first time diffusion in a realistic scale-free network. The results were highly surprising: In scale-free networks the epidemic threshold miraculously vanished! That is, even if a virus is not very contagious, it spreads and persists. Defying all wisdom accumulated during five decades of diffusion studies, viruses traveling in scale-free networks do not appear to notice any threshold. They are practically unstoppable.

The source of this highly unexpected behavior lies in the uneven topology of the Internet. Scale-free networks are dominated by hubs. Because each hub is linked to a very large number of other computers, it has a high chance of being infected by one of them. Once infected, a hub can pass the virus to all the other computers it is linked to. Thus, highly linked hubs offer a unique means by which viruses persist and spread. Whereas virulent species quickly reach all nodes in any network, in a scale-free environment their mildly contagious counterparts also have a good chance for survival.

These results are not limited to computer viruses. The models used by Pastor-Satorras and Vespignani, with some modifications, offer a simple description of the spread of ideas, innovations, and new products and the diffusion of infectious diseases. In a rough approximation, they capture the process that aids the spread of religions as well: Paul, a highly connected and mobile hub, helped the beliefs of early Christianity reach as many people as possible. Ideas and innovations diffuse from person to person along the links of the social web. Since the social network appears to have a scale-free topology, the anomalies observed in computer viruses should be present in these systems as well.

9.

Of the hundreds of social links each of us has, only a few are intimate enough to transmit a sexual disease. Therefore, AIDS advances on a very sparse subnet of our highly interlinked social web. Add to this the disease's relatively low contagiousness, and you find that the epidemic should have slowed and died out by now. Despite the odds, however, AIDS has already infected approximately 50 million people, and the numbers continue to rise. It is tempting to take the Trieste study at face value and attribute the rapid spread of the AIDS epidemic to the scale-free topology of the social network. But because not all social ties represent sexually active links, we need to ask, what is the topology of the sexual network that carries this deadly disease?

During a late November day in 2000, Carina Mood Roman, a Ph.D. student in sociology at the University of Stockholm, Sweden, was trying to make sense of an extremely skewed error plot she received while working on a class assignment. She had set out to predict the number of sexual partners of a group of Swedish subjects. Sexual mores in Sweden, one of the first countries to give legal rights to unmarried couples living together, are comparatively liberal. Sweden also prides itself on its remarkable and expansive health coverage and social services. As AIDS started to take its toll in northern Europe, Swedish researchers embarked on an extensive survey of sexual contacts, hoping to find the means to slow the epidemic.

Obtaining a map of the sex web, which links people via sexual relationships, is simply impossible. Would you be willing to give me the name of everybody with whom you have been intimately involved, knowing that I would then have to contact them all to sketch out their sexual links as well? Fortunately, we do not need a complete map of the sex web to decide whether it is scale-free or random. We need only measure the degree distribution by asking a representative subset of society how many sexual partners each has had. Not requiring our subjects to reveal the identity of their partners, we suddenly face a less challenging job. In 1996 Swedish scientists conducted thousands of interviews with a random sample of 4,781 individuals aged eighteen to seventy-four, collecting information regarding their sexual habits. With a response rate of 59 percent, they obtained the number of links for 2,810 nodes in the Swedish sex web.

Today students often are given the collected data to test various statistical methods. Roman had a copy of the data when she turned to her roommate, Fredrik Liljeros, for help in interpreting her error plot. Early in his sociology studies Liljeros was so favorably impressed during a series of lectures on mathematical sociology that he had devoted himself to the field, focusing on the evolution of social organizations. This research exposed him to a wide range of mathematical tools and concepts, including self-organization and power laws. Though typically Nordic in mien, when it comes to his passion, research, the twenty-something Liljeros does not share the stereotypical calm and reserved tone of his compatriots. "This looks like a power law!" he screamed to his roommate after spotting the plot on Roman's screen. Instead of helping her with her assignment, he asked for the data, and proceeded to verify his hunch. Next he e-mailed a copy to Luis Amaral at Boston University, with whom he had previously collaborated. Amaral had recently turned his attention to complex networks, authoring several seminal papers on modeling scale-free topology. He immediately saw that the data Liljeros e-mailed him contained the information key to answering our earlier question: What is the topology of the sex web?

Each study on our sexual habits faces severe memory biases: Men seem to remember more sexual partners than women do. Therefore, the subjects of the Swedish study first were asked to reveal how many sexual partners they'd had in the previous year only, in hopes the answer would be somewhat accurate. It was clear that their answers as to the number of partners they remembered having in their lifetime would be strongly affected by failing memories and expectations. Despite these potential biases, the results were consistent. They indicated that the majority of respondents had between one and ten sexual partners during their lifetime. Some, however, had dozens or more. A few had several hundred. The distribution followed a power law, regardless of whether one examined the one-year interval, considered all sexual partners, or focused only on either males or females. Taken together, the data offered striking evidence that the network of our sexual relationships has a scale-free topology, a conclusion reinforced by a subsequent study focusing on the American population.

Gaetan Dugas would seemingly hold the record with 250 sexual partners a year. But Wilt Chamberlain's claim that he'd had sex with a staggering 20,000 women clearly surpassed that measure. "Yes, that's correct, twenty thousand different ladies," he wrote. "At my age, that equals to having sex with 1.2 woman a day, every day, since I was fifteen years old." The NBA Hall of Famer's macho accounting made him a lighting rod for criticism by those offended by his promiscuity. The Stockholm-Boston collaboration, however, found that he is not that unique. The scale-free topology implies that, though most people have only a few sexual links, the web of sexual contacts is held together by a hierarchy of highly connected hubs. They are the Wilt Chamberlains and the Gaetan Dugas, collecting an astounding number of sexual partners.

In light of these results, the Trieste predictions offer a new perspective on the AIDS epidemic. The deadly virus must have followed the route already spotted in the spread of innovation and computer viruses: Hubs are among the first infected thanks to their numerous sexual contacts. Once infected, they quickly infect hundreds of others. If our sex web formed a homogeneous, random network, AIDS might have died out long ago. The scale-free topology at AIDS's disposal allowed the virus to spread and persist.

10.

When in 1997 we saw the first decline of AIDS deaths in the United States, we thought that the worst was over. We were wrong. Currently, every day 15,000 people are infected worldwide. The majority of them will die of the disease within a decade. If you are a fifteen-year-old in Botswana today, your risk of contracting and dying of AIDS during your lifetime is almost 90 percent. In fact we would be hard-pressed to pick a teenager from this or several other sub-Saharan countries who sooner or later will not be killed by the pandemic. This is despite the fact that several relatively effective treatments for AIDS are already on the market. To be sure, none of these treatments is a cure for the disease. But each does render it a chronic illness with which most patients can live almost indefinitely. The biggest problem is that these $15,000-a-year treatments are out of financial reach for most countries outside Europe and North America.

The crisis faced by Africa is the most severe. The problem is not only that most African countries cannot pay for the drugs. Even if drug prices were to drop, these nations lack the infrastructure to distribute and administer the treatment. At twenty, the AIDS epidemic has become a macabre celebrity. Through demonstrations and the aid of high-profile backers ranging from Bill Gates to any number of pop stars, AIDS activism has captured the spotlight, forcing the big pharmaceutical companies to deliver drugs at cost to poor nations. This is only the first step, however. It is clear that, despite the several-billion-dollars-strong international fund, there will not be enough money to buy treatments for everyone, even at cost. So who gets them?

Whereas the early spread of AIDS was attributed primarily to homosexual sex, today heterosexual sex is the leading means of transmission. As we've established, hubs play a key role in these processes. Their unique role suggests a bold but cruel solution: As long as resources are finite we should treat only the hubs. That is, when a treatment exists but there is not enough money to offer it to everybody who needs it, we should primarily give it to the hubs. This was the conclusion reached in two recent studies, one by Pastor-Satorras and Vespignani, the other by Zoltán Dezső, a graduate student in my research group. The results indicate that if we offer treatment for all nodes with a degree larger than a preselected value, no matter where we set the limit the epidemic threshold becomes finite. The more hubs we treat the larger the epidemic threshold and, thus, the higher the chance that the virus will die out.

The problem is that we do not know for sure who the hubs are. Therefore, Zoltán Dezső and I set out to address a more difficult question. While we do not know how to identify the hubs with a high degree of confidence, decades of research have produced numerous sociological methods for identifying high-risk groups, as well as individuals most likely to be the source of the epidemic, in a given community. Social status, age, occupation, and many other factors each play a role.

Therefore, with a certain probability, one can identify the hubs. Doubtless, many hubs will go undiscovered, while a few nonhubs will make the list. But we must ask whether such an imperfect method is useful. Since doing our best to identify the hubs is not enough, can we still restore the epidemic threshold? To answer this question we must assume that nodes are not treated randomly, but health organizations follow a biased policy that makes an individual with numerous sexual links more likely to be treated than those with only a few links. This stochastic approach allows us to compare those policies that are very effective in identifying and treating the highly connected nodes with those that distribute the treatment randomly. Zoltán Dezső undertook this comparison and we were surprised by the results. To be sure, each policy that continued to distribute the treatments randomly continued to have zero threshold and failed to stop the virus. But any policy that displayed bias toward the more connected nodes, even a small bias, restored the finite epidemic threshold. That is, even if we are not successful in finding all hubs, by trying to do so we can lower the rate at which the disease spreads.

Any selective policy raises important ethical questions. Indeed, our results indicate that, faced with limited resources, we would end up rewarding promiscuity: The more sexual partners an individual has had, the greater her or his likelihood of being picked for treatment. The better we are at selecting and treating the promiscuous individuals, the fewer people will be affected by the disease. Are we prepared to abandon the less connected patients for the benefit of the population at large? Are we ready to offer drugs to the more connected poor prostitutes than to the wealthier but sexually less connected middle class?

There is a solution that makes such moral debates academic: a vaccine. Currently, the world spends a mere $350 million annually on AIDS vaccine research. That number pales in comparison to the $3 billion plus annually spent on AIDS drugs in America and Europe or to the billion-plus dollar price tag for a single fighter plane. As we continue wrestling with our priorities, my feeling is that in the meantime we should do everything it takes to stop the spread of the disease, even if it requires rewarding promiscuity.

11.

Our understanding of hits and flops, epidemics and fads, has progressed considerably since the pioneering Iowa study. The last few decades have seen an incredible diversification of the subject. We have learned that studying the adoption of new crops can help us understand the spread of AIDS and the emergence of blockbusters. We have learned that, though randomness is involved in every diffusion process, the process follows laws that can be formulated in precise mathematical terms. And we have begun to understand the important role the social network plays in these processes.

Much has changed over the past five decades, however. The worldwide social network has imploded with the spread of high-speed communication devices, ranging from fax machines to e-mail, that bring and keep us together to a degree unprecedented in history. We feel a sense of urgency about understanding how this implosion affects the laws of diffusion. With the increasing threat of bioterrorism and with the steady spread of AIDS, there is a vital need to be able to predict and track deadly viruses in this increasingly mobile world, where infected individuals can hop on a plane and turn a local epidemic into a pandemic. In an increasingly computer-dependent world we have created a new breed of viruses that see no national boundaries. These cousins of Love Bug are more than mere nuisances. They represent a palpable threat to our security and way of life, easily capable of causing life-threatening emergencies. With their proliferation a new breed of epidemiologist has emerged, the computer security expert, who vigilantly monitors the health of our online universe.

Innovations and biological or computer viruses spread across inhomogeneous networks where hubs run the show. The implications of the Trieste study are that we are in for more surprises when it comes to how much we know about spreading and diffusion. I believe that the results obtained so far represent only the tip of the iceberg. Whereas spreading and diffusion have universal properties, individual systems have unique features that are often as important as some of the generic laws. We would be kidding ourselves if we believed that modeling computer viruses would give us a good picture of the AIDS epidemic. To be able to make detailed predictions on the disease's spread, the models should include many details that are specific to the pandemic. That is still a distant dream. But understanding the fundamental laws that govern spreading and diffusion is the key prerequisite for success. The recent breakthroughs in these directions offer a strong impetus to revisit problems ranging from marketing to the spread of influenza and to critically inspect the inherent assumptions. As we follow this path, I am convinced that many more surprises and potential breakthroughs will surface.

The recent paradigm changes in diffusion and epidemics studies were possible thanks to the wealth of data offered by the Internet, one of the most charted networks. The Internet helped us discover scale-free networks in the first place. The viruses navigating it provided the insights and the necessary data that made the Trieste study possible, uncovering the threshold-free nature of some epidemics. The understanding they offered has prompted us to revisit everything from fads to the AIDS pandemic. Let us step back now and take a look at the entangled medium that made all these discoveries possible and chart the network behind it.

THE ELEVENTH LINK

The Awakening Internet

WHEN PAUL BARAN REGISTERED a week late for his first computer science class at University of Pennsylvania, he knew that he had already missed the first lecture, but he was not too worried. Not much is done in the first class anyway. So he showed up for the second class, on Boolean algebra, the mathematics behind computer logic. As he recalls, "The instructor went up to the blackboard and wrote '1 + 1 = 0.' I looked around the room waiting for someone to correct his atrocious arithmetic. No one did. So I figured that I may be missing something here, and I didn't go back." Yet, he did revisit the subject ten years later, on his fourth job after graduation. This time he faced a different problem: He was way too early.

Barely thirty and only a few months into his new job at RAND Corporation, Baran was given the prodigious task of developing a communication system that would survive a nuclear attack. In 1959 the possibility of a Soviet nuclear warhead's falling from the sky was not mere science fiction but an appropriately feared potential war scenario. Baran's employer, a California think tank founded in 1946 to provide the intellectual know-how for the military's nuclear buildup, had considerable expertise in developing war scenarios and potential disaster outcomes. Such grim tasks as foreseeing and detailing the death of millions from a nuclear attack were never a source of good press, often tarring the company with Dr. Strangelove's brush. Baran's assignment, to develop a survivable communicator system, was par for the course at RAND. Baran took his job seriously, and in a twelve-volume series of RAND Memorandums he meticulously described the vulnerabilities of the existing communication infrastructure and proposed a better one—the Internet.

Baran saw the vulnerability of the command system of the 1950s hidden in the topology of the existing communication network. Since a nuclear strike handicaps all equipment within the range of detonation, he wanted to design a system whose users outside of this range would not lose contact with one another. Inspecting the communication systems of that time, he saw three types of networks (see Figure 11.1). Baran discarded the starlike topology, concluding that "the centralized network is obviously vulnerable as destruction of a single central node destroys communication between the end stations." Baran saw the current system as a "hierarchical structure of a set of stars connected in the form of a larger star," offering an early description of a scale-free network. With incredible insight, he found this topology too centralized to be viable under attack. In Baran's mind the ideal survivable architecture was a distributed meshlike network, similar to a highway system, redundant enough so that even if some nodes went down, alternative paths maintained the connection between the rest of the nodes.

An enduring myth alleges that the Internet was designed to survive a Soviet nuclear strike. It is true that Baran's main motivation was to design a system that could not be taken out by the Soviet nuclear arsenal. But in the long run his ideas and innovations were all but ignored by the military. As a result the topology of today's Internet has little to do with his vision. Yet the topological change advocated by Baran was not the reason everyone from the military to industry vehemently opposed his design. The objection was to his proposal to break the messages into small packets of uniform size capable of traveling independently of one another along the network. This could not be achieved with the existing analog communication system. Thus he advocated a switch to a digital system. This step was too difficult for AT&T, the communication monopoly of his time, to absorb. Therefore, AT&T's Jack Osterman quashed Baran's vision when he declared, "First, it can't

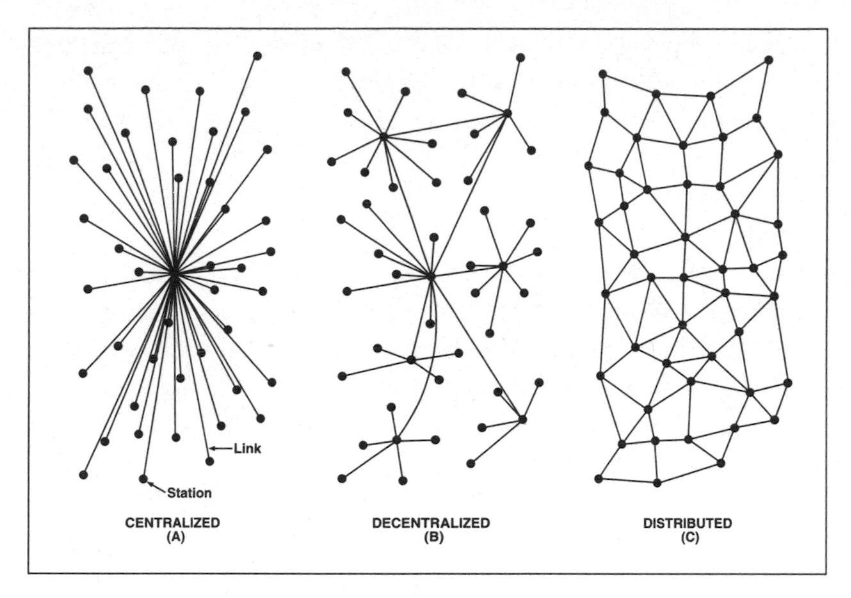

Figure 11.1 Paul Baran's Networks. *In 1964, Paul Baran began thinking about the optimal structure of the Internet. He suggested that there were three possible architectures for such a network—centralized, decentralized, and distributed—and warned that both the centralized and decentralized structures that dominated communications systems of the time were too vulnerable to attack. Instead, he proposed that the Internet should be designed to have a distributed, mesh-like architecture. (Reproduced with permission of Paul Baran.)*

possibly work, and if it did, damned if we are going to allow the creation of a competition to ourselves." Baran's ideas, defeated at every step by industry and the military, were rediscovered only years later, when the Advanced Research Projects Agency, not aware of his results, independently constructed the same vision. By that time, however, the Internet was well along its course of development.

Understanding the topology of the Internet is a prerequisite for designing tools and services that offer a fast and reliable communication infrastructure. Though human made, the Internet is not centrally designed. Structurally, the Internet is closer to an ecosystem than to a Swiss watch. Therefore, understanding the Internet is not only an engineering or a mathematical problem. In important ways, historical forces shaped its topology. A tangled tale of converging ideas and competing motivations left their mark on the Internet's structure, creating a jumbled information mass for historians and computer scientists to unravel.

1.

The Advanced Research Projects Agency, or ARPA, was President Eisenhower's answer to the Soviets' launching of the first Sputnik satellite. Originally ARPA had sweeping control of the most advanced military research and development projects, in particular the antimissile and satellite programs. It lost its muscle, however, after NASA took over the space program.

Struggling for a mission, ARPA reinvented itself to coordinate long-range research relevant to the military, in contrast with the immediate developmental projects that different military agencies were handling themselves. The Internet entered the picture around 1965 or 1966, when Bob Taylor, the director of ARPA's computing program, suddenly became concerned with a huge waste of federal resources he had just discovered.

In the 1960s, ARPA was already funding computer research in a big way. This indeed required considerable investment—with the PC revolution decades away, computers cost anywhere from half a million to several million dollars. ARPA had several such monsters in its research portfolio, hosted by research labs around the country. The problem was that even computers in the same room could not talk to each other. Tapping into the computing power stored at other ARPA-supported sites was out of the question. Bob Taylor had a brilliant idea: To stop this waste, why not link these incompatible machines somehow? In February 1966, after presenting his vision to Charlie Herzfeld, ARPA's director, he walked away with a fresh million in his budget and a new sense of mission.

The idea of connecting computers also occurred to Donald Davies, director of computer science at Britain's National Physical Laboratory in Teddington, a town within commuting distance of London. Working hard to turn his idea into reality, Davies reinvented packets and packet switching well before learning of Baran's preexisting work. His group presented these concepts at a 1967 symposium in Gatlinburg, Texas, introducing his and Baran's ideas to the ARPA-supported research group. It suddenly became clear to everyone that packet switching over faster lines was the technology required to create a truly efficient communication network. Finally Baran's decade-old vision began to materialize. And so the network that today we call the Internet was born.

The word *Internet* is often used to describe everything related to our online universe, including computers, routers, optical cables, and even the World Wide Web. Here we will use the word to refer only to the physical infrastructure connecting computers. The Internet is a network of routers that communicate with each other through protocols envisioned by Paul Baran and made possible thanks to ARPA's deep pockets. Ironically, the principles directing today's Internet match Baran's original vision in every respect except the guiding principle that motivated his work: undercutting vulnerability to attacks. Baran's distributed highwaylike network could have become a reality only if the Internet had continued to be regulated and maintained by the military. The Internet, however, took on a life of its own.

2.

In the computer science community Bill "Ches" Cheswick, a researcher at a Lucent/Bell Labs' spin-off called Lumeta, is best known for his work on firewalls and computer security. But the public increasingly recognizes him for the colorful Internet maps he and Hal Burch, also at Lumeta, produce and sell through Peacockmaps.com. The millennium map, depicting the Internet's topology on January 1, 2000, shows a dense, entangled forest of routers and links, a network of considerable beauty. Its complexity is matched perhaps only by the human brain. There is an important difference between the two, however. Whereas the human brain's size has been stagnating for centuries, the Internet continues to grow exponentially, without any sign of slowing down.

Cheswick is far from being a lone scientist with artistic aspirations. He is in illustrious company. DARPA, the successor to ARPA, is currently spending millions of dollars on research groups around the United States to do just what Cheswick is doing: map the Internet. The most visible of these projects is the Cooperative Association for Internet Data Analysis, or CAIDA, an Internet tomography collaboration hosted by the University of California at San Diego, whose main goal is to monitor just about every characteristic of the Internet from traffic to topology. Across the Atlantic but only a click away, Martin Dodge, a researcher at the Center for Advanced Spatial Analysis at University College London, hosts Cybermaps.com, a colorful Website collecting a stunning body of maps visualizing the Internet.

Would it ever occur to you to meticulously draw a map of your watch, the Pentium chip in your computer, or the car you drive everyday to work? Hardly. If you really want to know what is under the hood, you could contact the manufacturer for the car's blueprint. Engineers prepare hundreds of maps before building each watch, chip, or car, detailing not only every component, but the location of and the relationship between each piece, as well. But today, when the Internet is the workhorse of the American economy, we still do not have a detailed map of it. Since the National Science Foundation relinquished its stewardship of the Internet in early 1995, no central authority has controlled or documented its growth and design.

Today the Internet evolves based on local, distributed decisions on an "as needed" basis. Everyone, from corporations to educational institutions, adds nodes and links without needing permission from a central authority. There is no single network either. Independent but interlinked networks coexist and operate, going by such names as WNET, vBNS, or Abilene.

You would think there was someone out there who, if necessary, could shut the whole thing down. Wrong. While you could persuade an institution to close down the portion of the network under its authority, no single company or person controls more than a negligible fraction of the whole Internet. The underlying network has become so distributed, decentralized, and locally guarded that even such an ordinary task as getting a central map of it has become virtually impossible.

3.

There are important practical reasons for seeking a global Internet map. Without knowing the Internet's topology it is impossible to design better tools and services. The current Internet protocols were developed with a small network and 1970s technologies and needs in mind. As the network grew and new applications emerged these protocols have often fallen short of our desires. Indeed, most of today's use of the Internet was unimaginable by those who designed the basic infrastructure, which is still in place. For example, e-mail was born when an adventurous hacker, Rag Tomlinson, working at BBN, a small consulting firm in Cambridge, Massachusetts, figured out how to modify the file transfer protocols to carry mail messages. For a long time Tomlinson kept quiet about his breakthrough. When he first showed it to one of his colleagues, he warned him, "Don't tell anyone! This isn't what we're supposed to be working on." E-mail leaked out, however, and became one of the dominant applications of the early Internet.

The same is true of the World Wide Web. The infrastructure was never prepared for it. It is an excellent example of a "success disaster," the design of a new function that escapes into the real world and multiplies at an unseen rate before the design is fully in place. Today the Internet is used almost exclusively for accessing the World Wide Web and e-mail. Had its original creators foreseen this, they would have designed a very different infrastructure, resulting in a much smoother experience. Instead we find ourselves locked into a technology that adapts only with great difficulty to the booming diversity and demand imposed by the increasingly creative use of the Internet.

Until the mid-nineties all research concentrated on designing new protocols and components. Lately, however, an increasing number of researchers are asking an unexpected question: What exactly did we create? While entirely of human design, the Internet now lives a life of its own. It has all the characteristics of a complex evolving system, making it more similar to a cell than to a computer chip. Many diverse components, developed separately, contribute to the functioning of a system that is far more than the sum of its parts. Therefore, Internet researchers are increasingly morphing from designers into explorers. They are like biologists or ecologists who are faced with an incredibly complex system that, for all practical purposes, exists independently of them. The mystery is a bit deeper than that, however. While biologists have spent decades figuring out what proteins look like and how they interact with each other, all details regarding the Internet's components are fully available to the Internet tomographer. What neither computer scientists nor biologists know is how the large-scale structure emerges once we put the pieces together.

4.

Vern Paxon and Sally Floyd, computer scientists at the International Computer Science Institute Center for Internet Research in Berkeley, California, in an influential and much quoted 1997 paper, identified our limited knowledge of the network topology as the main obstacle toward a better understanding of the Internet as a whole. Two years later three Greek computer scientist brothers, Michalis Faloutsos of the University of California–Riverside, Petros Faloutsos of the University of Toronto, and Christos Faloutsos of Carnegie Mellon University, made a surprising discovery. They found that the connectivity distribution of the Internet routers follows a power law. In their seminar paper "On Power-Law Relationship of the Internet Topology" they showed that the Internet, a collection of routers linked by various physical lines, is a scale-free network. Their discovery had a simple message that quickly penetrated the research community: All tools used to model the structure of the Internet before 1999, based on ideas rooted in random networks, were simply wrong.

The Faloutsos brothers were unaware of the parallel discoveries of power laws in the World Wide Web topology. Combined with these developments their finding acquired a new meaning, removing the Internet from the world of random networks and dropping it into the colorful zoo of scale-free topologies. This was rather unexpected. After all, the Internet is comprised of physical lines and routers. It is all hardware. How could these costly and heavy copper and optical connections follow the same rules as humans do when establishing their weightless social links or adding URLs to their Webpage?

5.

In October 1969 Charley Kline was asked to arrange the first computer-to-computer message through an ordinary telephone line. Working as a programmer in the UCLA lab of Leonard Kleinrock, he was part of a project attempting to connect to the only other existing Internet node located at Stanford University. After establishing the connection, Kline started by typing "login." He typed *l* and got the echo from Stanford confirming that the letter had been received. He proceeded with *o* and again received the appropriate echo. Then he ventured to *g*. However, that was too much for the young system to absorb, and the computer crashed, killing the connection as well.

The connection was quickly reestablished, and after the UCLA and Stanford nodes were firmly in place many others joined in. According to John Naughton, author of *A Brief History of the Future*, the University of California–Santa Barbara and the University of Utah got the third and the fourth nodes in November and December 1969, respectively. The fifth was delivered to BBN, a Massachusetts consulting firm, early in 1970, together with the first cross-country circuit—a second line connecting the machines in Los Angeles to BBN's in Boston. By the summer of 1970, nodes six, seven, eight, and nine had been installed at MIT, RAND, System Development Corporation, and Harvard. By the end of 1971 the Internet consisted of fifteen nodes; by the end of 1972 it had thirty-seven. As Naughton puts it, "The system was beginning to spread its wings—or, if you were of a suspicious turn of mind, its tentacles."

As you may have noticed, the Internet follows the classical scenario of a growing network. Today, two decades later, it continues to expand node by node—the first and necessary condition for the emergence of a scale-free topology. Preferential attachment, the second condition, is more subtle, however. Why would anyone link his or her computer to any router other than the nearest one? After all, laying down a longer cable is more expensive.

It turns out that the length of cable is not the limiting factor determining the growth or stagnation of the Internet. When an institution decides to link its computers to the Internet, it has only one parameter in mind: cost of communication. Regarding bandwidth, the measure of how many bits a connection can carry each second, the closest node is often not the best choice. Going a few extra miles could provide access to faster routers.

Routers offering more bandwidth likely have more links as well. Thus, while shopping for a good place to link, network engineers inevitably gravitate toward the more heavily connected access points. This simple effect is a possible source of preferential attachment. We do not know for sure whether it is the only one, but preferential attachment is unquestionably present on the Internet. This was first demonstrated by Soon-Hyung Yook and Hawoong Jeong, both working in my research group, when they compared Internet maps recorded at several months' time intervals. Charting how the Internet grows node by node they found quantitative evidence that nodes rich in links acquire more links than nodes with a few links only.

Growth and preferential attachment should be sufficient to explain the scale-free topology discovered by the Faloutsos brothers. On the Internet things are a bit more complicated, however. While not the primary consideration, distance does matter. Undeniably, it is more expensive to lay down two miles of optical cable than half a mile. We must also take into consideration that nodes do not appear randomly across the map. Routers are added where there is a demand for them, and demand depends on the number of people wanting to use the Internet. Thus there is a strong correlation between population density and the density of the Internet nodes. The distribution of routers on the map of North America forms a fractal set, a self-similar mathematical object discovered in the 1970s by Benoit Mandelbrot. Therefore, when trying to model the Internet, we must simultaneously acknowledge the interplay of growth, preferential attachment, distance dependence, and an underlying fractal structure.

Each of these forces alone, if taken to the extreme, could destroy the scale-free topology. For example, if the length of the wire were the main consideration when deciding where to link, the resulting network would have an exponential degree distribution, developing a topology very similar to the highway system. But the amazing thing is that these coexisting mechanisms delicately balance each other, maintaining a scale-free Internet. This very balance of power is the Internet's own Achilles' heel.

6.

MAI Network Services, a small Internet service provider headquartered in McLean, Virginia, owns several high speed Internet routers linked to the giant networks owned by Sprint and UUNet. On the morning of Friday, April 25, 1997, MAI released a routing table update for its routers. Routers shepherd packets they receive toward their destination by matching the address on the packet with a routing table. These routing tables are the roadmaps of the Internet. As the network topology is constantly changing, the routing tables are also periodically updated. At 8:30 A.M. MAI broadcast the updated routing information to its own routers. Because of an incorrect configuration, the update did not stop at the borders of MAI but escaped and rewrote the routing tables of a large number of routers owned by Sprint and UUNet. It instructed them to send all traffic to several MAI routers.

It was like watching water burst from a broken dam, destroying everything in its path. MAI watched in horror as all Internet traffic was suddenly redirected towards it. Because it never had the capacity to handle even a fraction of this flood, MAI turned into a black hole, absorbing packages at an incredible rate. Forty-five minutes later the company was forced to shut itself down to stop the damage. In the meantime Internet providers helplessly watched all their traffic get sucked into the black hole created by the faulty reconfiguration. Sprint recovered only after it manually changed all the routing tables it owned, as did many of the big and small Internet providers affected by the problem.

Thanks to the quick resolution and the relative youth of the Internet, the world paid little attention to the event. However, it offered a vivid demonstration of the speed by which errors propagate on the Net: Within minutes of its release the misconfigured routing table was part of several large networks, triggering a classic example of a cascading failure.

Paul Baran had a very specific threat in mind when he designed the prototype of the Internet. He anticipated Soviet nuclear warheads hitting intelligence and military headquarters, potentially leading to complete information and communication loss. Neither he nor the early Internet pioneers considered the possibility that one day people from any country in the world could have access to the infrastructure. For many years the United States resisted sharing the technology with countries deemed nonfriendly. I experienced that myself, as the much hated COCOM list officially excluded Hungary from the Internet until the fall of the Berlin Wall. The Internet was too contagious to be halted by such artificial barriers, however. Thanks to the ingenuity of local system managers, many eastern European universities had been regularly communicating via e-mail with their Western colleagues well before the restrictions were lifted. Today virtually every country on Earth is connected to the Internet. This open access policy brought along unexpected dangers and vulnerabilities as well, increasingly threatening our interlinked world.

One of the United States' busiest nodes, owned by AT&T, is a highly guarded subterranean facility in Schaumburg, Illinois, a Chicago suburb. This and several similarly well protected key nodes offer a false sense of security that the Internet cannot be broken by intentional attacks. The increasingly understood interplay between the network architecture and the protocols presents a different picture, however. A few well-trained crackers could destroy the net in thirty minutes from anywhere in the world. There are many ways to accomplish this, from breaking into the computers running several key routers to launching denial-of-service attacks against the busiest nodes.

The Code Red worm, which spread like a virus and infected hundreds of thousands of computers worldwide in the summer of 2001, is a good example of a technology that could achieve just that destruction. At first it appeared to be a harmless virus, since it did not damage its host. But after sitting dormant for days, it suddenly turned all infected computers into zombies, simultaneously throwing traffic at whitehouse.gov. Code Red was only a proof-of-principle demonstration of what automated viruses could achieve. More sophisticated versions could result in unparalleled damage. Disabling a few major nodes would not be sufficient to break the network into pieces, but the cascading failure of other routers resulting from the redirection of traffic to smaller nodes would finish the job.

Most crackers or hackers with the know-how would have no interest in taking the whole Internet out. A successful attack would take away their favorite toy, denying them access to the Net, as well. So a large-scale action taking on the entire Internet would never be the work of true hackers. But it could easily be the goal of rogue nations and terrorists. Understanding the Internet's topology will help us protect it.

7.

On August 30, 2001, National Public Radio aired a five-minute segment about our latest research, published the same day by the British journal *Nature*. It was not the first time that our work had been featured in the media. But the next morning, staring in disbelief at the project's Website counter, which had registered over 10,000 hits overnight, I realized things were a bit different this time. My e-mailbox was crowded with uncountable messages. Most were positive. Some, however, were rather scary. "Stay the hell out of my computer!" wrote a senior officer in a company developing deterrence programs. "I'd hate to see another Eastern European CompSci person tossed in jail by the US Federal Government," concluded another less than friendly note, reminding me of the recent arrest of a Russian hacker by U.S. authorities. "[I] request that you assure us that no computers on our networks have been, or are currently being, targeted by this program," wrote the CEO of a company in

Norway. "I remind you that any unauthorized use of resources located at these IP addresses is illegal and may result in legal action and demands for compensation." How could a research paper intended for an academic audience and published by one of the most prestigious scientific journals create such a fierce and immediate reaction?

James McAdams, head of the Department of Government and International Studies at Notre Dame, had a great idea in early 2000. He assembled seven professors from all different departments, including economics, physics, law, chemical engineering, computer science, and Asian languages, to discuss in an informal setting the impact of the Internet on everything from democracy to teaching. Meeting once a month for lunch or breakfast, we took turns suggesting discussion topics and assigning reading materials, covering issues from cyberlaw to social movements on the Web. During one such breakfast meeting computer scientist Jay Brockman mentioned that the Web is a computer, metaphorically speaking. His comment left me puzzled. To be sure, the Internet is comprised of computers that can exchange Webpages and e-mail messages. But this limited, user-driven communication does not yet make the World Wide Web a single computer.

Could we do something to change this? Could we make computers drive each other's activity? To get started, could I force any computer out there to do computation on my behalf? Now this was an interesting question that I was willing to entertain. We ended up forming a tiny research group to try to address it. Brockman and I were soon joined by Vincent Freeh, an expert on Internet protocols, and my longtime collaborator Hawoong Jeong. After many discussions and tutorials on how computers communicate, a simple but controversial idea emerged: *parasitic computing.*

Sending a message through the Internet is a sophisticated process regulated by layers of complex protocols. For example, when you click on a URL to view a Webpage, your request is broken into small packets that are then carried to the computer owning the Webpage. There the request is reconstructed and interpreted, prompting the distant computer to send you the requested Web document. Therefore, such a seemingly simple task as clicking on a URL involves a significant amount of computation along the way. Parasitic computing exploits this setup by forcing computers to perform computation at the command of a master host by merely engaging the computers in communication. To achieve this we disguised complex computational problems as legitimate Internet requests. When a computer received a packet, it performed a routine check to ensure that the packet had not been corrupted during its journey. While doing the math, it solved a problem of interest to us, encoded into the packet.

Our implementation of parasitic computing demonstrated that we *can* enslave computers located thousands of miles away, forcing them to perform computation on our behalf. This fundamental vulnerability of the Internet raised a barrage of computational, ethical, and legal questions. What if someone improves the method, making it efficient, and starts using it on a grand scale? Who owns the resources that are made available to anyone through the Internet? Could this mark the birth of the Internet computer? Will there be a new intelligent being at the end of this road?

Taken to an extreme, parasitic computing suggests that in the future computers could swap information and services on an as-needed basis. Right now communication within a chip is orders of magnitude faster than communication across the Internet. With broadband communication on its way, the gap will shrink. Soon it will start making perfect sense to ask other computers to chip in their unused resources to solve complex problems that cannot be addressed by a single computer or research group. On a smaller scale this possibility has already been exploited by SETI@home, a Berkeley-based project that harbors the unused time of millions of PCs to search for extraterrestrial intelligence.

The SETI model requires your voluntary collaboration. Most of us are just simply too lazy to go along. If, however, protocols allowing service and information swapping become the norm, vast unused resources could be tapped. Along the way the Internet might become independent of human supervision, since it can shepherd most of the information and resources it needs to solve specific problems. This could have unforeseen impact on the Internet's topology as well, giving self-organization an even bigger role. I can imagine a time when, after getting an answer to a question from your Web browser, neither you nor your computer will know for sure where it came from. After all, do you know where the letter *A* is stored in your brain?

8.

Our skin is a unique piece of engineering. It has the ability to measure and sense changes in temperature and movement of air; it can size up objects and identify their make. It achieves all of this with the help of a huge number of tiny integrated chemical sensors that talk to each other through the nervous system. As Neil Gross pointed out in *Business-Week*, a skin of similar sensitivity is enfolding the earth right now. Millions of measuring devices, including cameras, microphones, thermostats and temperature gauges, light and traffic sensors, and pollution detectors, are popping up everywhere, feeding information into increasingly fast and sophisticated computers. Experts predict that by 2010 there will be around 10,000 telemetric devices for each human on the planet. This number is not particularly significant in and of itself—we've had sensors for a long time, ranging from surveillance cameras in supermarkets to car detectors buried in the pavement at traffic signals that switch the lights at the intersection. What is changing is that for the first time these various sensors are feeding information into a single integrated system. There will soon be over 3 billion Internet-connected cell phones and close to 16 billion Internet-connected computers embedded in everything from toasters to fashion designs. The tiny sensors of this planetary skin will spy on everything from the environment to our highways and bodies. Most importantly, however, they are all connected. Our planet is evolving into a single vast computer made of billions of interconnected processors and sensors. The question being asked by many is, when will this computer become self-aware? When will a thinking machine, orders of magnitude faster than a human brain, emerge spontaneously from billions of interconnected modules?

It is impossible to predict when the Internet will become self-aware, but clearly it already lives a life of its own. It grows and evolves at an unparalleled rate while following the same laws that nature uses to spin its own webs. Indeed, it shows many similarities to real organisms. Just like the millions of reactions taking place in a cell, terabytes of information are passed along its links every day. The surprising thing is that some of this information is very difficult to find. That brings us to yet another network: the World Wide Web.

THE TWELFTH LINK

The Fragmented Web

SCIENCE FICTION WRITERS and visionaries, whose books I consumed as a child, made me believe that by the turn of the century human-looking robots would handle all mundane tasks. Yet we entered the new millennium without such humble servants having appeared on the scene. Or perhaps the robots have arrived quietly. They do not have the shining golden exterior of the always worried C-3PO, nor can they produce the joyful whistle of R2-D2. They wisely avoid sharing the crowded Euclidean space with us, where real estate is at a premium. The robots of the twenty-first century are invisible and immaterial. They have taken up residence in the virtual world, which allows them to hop with enviable ease from continent to continent. Staring at your computer screen won't reveal these robots. But if you take the time to inspect carefully your computer's log files, which keep detailed records of who has visited your Webpage, you can catch them in action. You will see them tirelessly performing one of the most thankless and boring jobs humanity has ever designed: reading and indexing millions of Webpages.

Designed for speed and efficiency, these robots—the sports cars of the Web—rapidly sweep along the links, sniffing out just about everything in their paths. While these road warriors overshadowed the little beetle Hawoong Jeong built to map the Web, I was truly proud of it. It was like the first used car one could finally afford. And it crashed just about every other day, often getting into trouble by inadvertently carrying home Webpages protected by robot exclusion files.

It soon became clear that mapping the whole Web was a dream beyond the capabilities of our little engine. But sneaking and often stalled, it managed to carry home about 300,000 Webpages, enough to discover that there are scale-free networks out there. We shut it down at that point—perhaps a bit too early. Had we let it go further and allowed it to bring home a larger sample of the Web, we might have discovered other features of complex networks that were not so evident from our smaller sample. Search engines do see a much larger portion of the Web than we did during our experiments. Researchers studying these huge samples have made some fascinating discoveries. They have found that the Web is fragmented into continents and communities, limiting and determining our behavior in the online universe. Paradoxically, they have also told us that there is terra incognita out there, whole continents of the Web never visited or seen by robots. Most important, we learned that the structure of the World Wide Web has an impact on everything from surfing to democracy.

1.

A few years ago we thought we knew everything there was to know about the Web. Comments like "If you can't find it using AltaVista, it's probably not out there" or "HotBot is the first search robot capable of indexing and searching the *entire* Web" were routine. We trusted the search engines to cover and deliver the Web to us. This suddenly changed in April 1998. "We prefer to index quality sites instead of a greater quantity of sites" was the new spin from the spokesman of a major search engine. Others went even further, claiming that "many pages are not worth indexing." What happened? This sudden mood shift was provoked by a research paper published on April 3, 1998, in the journal *Science*. Its three pages completely changed our perceptions about the accessibility of information stored on the Web.

Steve Lawrence and Lee Giles never planned to undermine the credibility of search engines. Working at the NEC Research Institute in Princeton, New Jersey, they were interested in machine learning, a booming subfield of computer science. They built a meta–search engine, a robot called *Inquirus* that could inquire at each major search engine for documents matching a given query. Halfway through they realized that their robot could do more than it was originally designed for: It could help them estimate the size of the Web.

Inquirus asked several search engines to list all documents containing a given word, for example, *crystal*. If each search engine visits and indexes the full Web, it must return the same list of documents. In reality the lists returned by different search engines are rarely identical. There is always significant overlap, however. For example, of the 1,000 documents containing the world *crystal* found by AltaVista, 343 were on HotBot's list as well. Dividing the number of overlapping documents by the number of documents returned by AltaVista gives the fraction of the Web covered by HotBot. Since HotBot reportedly indexed 110 million pages in December 1997, the NEC group estimated that the World Wide Web had approximately 110/0.343 million, or about 320 million documents at the same time. Today this number may not seem that large. In 1997, however, this was at least twice the current best guess of the Web's size.

Before 1998 we believed everything the search engines told us about the size of the Web. After all, they should know. Lawrence and Giles's landmark study turned the Web into a target of scientific inquiry—one that could and must be studied using systematic and reproducible methods. But their findings about the search engines' ability to map the Web offered us little to cheer about.

2.

According to the NEC study in 1997 HotBot collected the largest number of documents, earning the distinction of being the search engine with the largest coverage. This was great news for the company. David Pritchard, marketing director for HotBot, proudly acknowledged this:

"We're the largest index out there—there are no surprises for us in this report." Well, there were some. The bad news was that HotBot covered only 34 percent of the full Web. That is, 66 percent of all Webpages were unseen by it. AltaVista, the most popular search engine at that time, was second on the list because its robots sniffed out only 28 percent. Some search engines, such as Lycos, had captured as little as 2 percent. Their reaction was predictable: "Quite frankly, I don't give these kinds of reports a lot of credence. Our focus is not on quantity, it's on quality," said Rajive Mathur, senior product manager at Lycos Inc.

One would think that the NEC study would have motivated the search engines to increase their coverage. It didn't. A year later, in February 1999, Lawrence and Giles repeated their measurements and found that the size of the Web had more than doubled, swelling to 800 million documents, but the search engines had not kept up with this growth. In fact, their coverage had further deteriorated. This time Northern Light was the leader, covering a mere 16 percent of the World Wide Web. HotBot and AltaVista had lost significant ground: Their coverage decreased to 11 and 15 percent, respectively. Google indexed only 7.8 percent of the estimated 800 million pages out there. Taken together, in 1999 the search engines covered about 40 percent of the full Web. That means that six out of ten pages relevant to your query would never be returned by *any* search engine. Simply, they would have never seen it.

Eventually the NEC study did ignite a fierce competition among the search engines. Size suddenly mattered. A fight for dominance developed between AltaVista and the new search engine run by FAST, whose address, alltheweb.com, leaves little room for ambiguity regarding the company's goal. In January 2000 alltheweb.com broke the 300-million-page mark. AltaVista followed shortly. By June 2000 the new kid on the block, Google, had become a serious contender, breaking the 500-million mark. Inktomi soon matched that, and so did yet another newcomer, WebTop.com. In June 2001 Google hit a new record, reaching for the first time the magic 1-billion-document coverage mark.

As of now Google maintains the lead. Alltheweb.com, pursuing its dream to eventually map out the full Web, is second with over 600 million documents, followed by AltaVista with 550 million. The search engines are doing better and better. This is great news. There is one problem, however: The Web is growing even faster.

Most search engines do not even try to reach the full Web. The reason is simple: The search engine with the most documents is not necessarily the best one. To be sure, if you are looking for difficult-to-find information, the engine with the larger coverage is your best bet. But when it comes to popular topics, a larger index does not necessarily offer better results. Most of us are already overwhelmed by the thousands of hits search engines return for simple queries. The last thing we want is to see millions more. Therefore, beyond a certain point it is more profitable to enhance the algorithm that selects the *best* page from the search engine's already enormous database than to go deeper into the Web.

When it comes to surfing the Web, either by individuals or robots, economic incentives (or their absence) are not the only limitations. The topology of the Web limits our ability to see everything out there. The World Wide Web is a scale-free network, dominated by hubs and nodes with a very large number of links. But, as we will see next, this large-scale topology coexists with numerous small-scale structures that severely limit how much we can explore simply by clicking our way along the links.

3.

Despite the billion documents on the Web, nineteen degrees of separation suggests that the Web is easily navigable. Big yet small. But the small world behind the Web is a bit misleading. To be sure, if there is a path between two documents, that path is typically short. But in reality not all pages can be connected to each other. Starting from any page, we can reach only about 24 percent of all documents. The rest are invisible to us, unreachable by surfing.

This is a consequence of the fact that for various technical reasons the links of the Web are directed. In other words, along a given URL we can travel only in one direction. If there is no direct link between

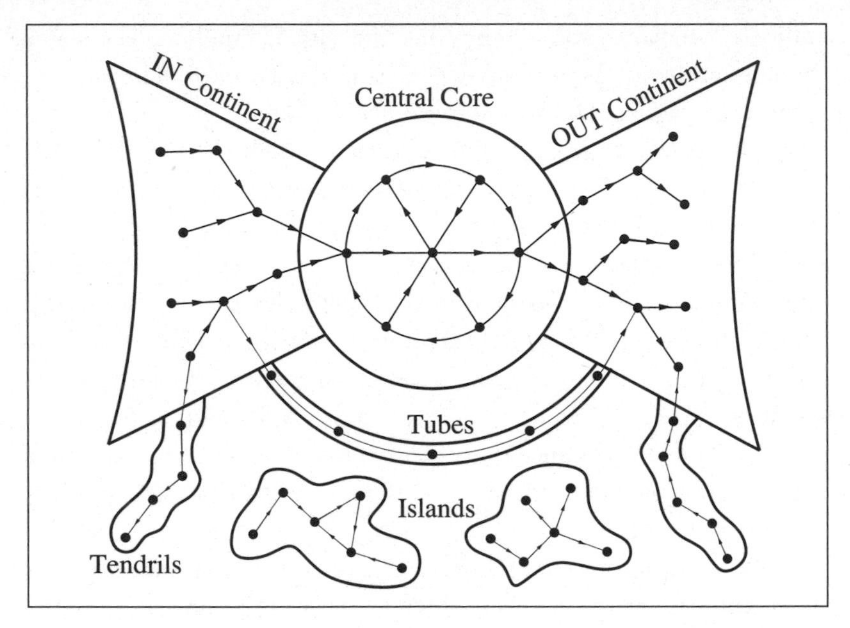

Figure 12.1 The Continents of a Directed Network. *Directed networks such as the World Wide Web naturally break down into several easily identifiable continents. In the central core each node can be reached from every other node. Nodes in the IN continent are arranged such that following the links eventually brings you back to the central core, but starting from the core doesn't allow you to return to the IN continent. In contrast, all nodes of the OUT continent can be reached from the core, but once you've arrived, there are no links taking you back to the core. Finally, tubes directly connect the IN to the OUT continent; some nodes form tendrils, attached only to the IN and OUT continents; and a few nodes form isolated islands that can't be accessed from the rest of the nodes.*

two nodes in a directed network, you can connect them through other nodes: For example, if you want to go from A to D, you can start from node A, then go to node B, which has a link to node C, which points to D. But you can't make a round-trip. In a nondirected network, where you can follow a link in both directions, an A → B → C → D path implies that the shortest path from D to A is the reverse one, D → C → B → A. In a directed network, however, there is no guarantee that the inverse path exists. Most likely you would have to follow a different route back: From D you might need to visit dozens of intermediate nodes before getting back to A. The Web is full of such disjointed directed paths. They fundamentally determine the Web's navigability.

Directed networks do not represent a fundamentally new class of networks: Whether the network is scale-free or random, the links can be either directed or nondirected. So far we have dealt with mostly nondirected links. Indeed, most webs, ranging from social to protein interaction networks, are nondirected. But some networks, ranging from the World Wide Web to food webs, have directed links. This directedness has consequences for the network's topology. In the context of the World Wide Web these consequences were first addressed by Andrei Broder, from AltaVista, and his collaborators from IBM and Compaq. They studied a sample of 200 million nodes, close to a fifth of all Webpages in existence in 1999. Their measurements indicated that the most important consequence of directedness is that the Web does not form a single homogeneous network. Rather, it is broken into four major continents (see Figure 12.1), each forcing us to obey different traffic rules when we want to navigate them.

The first of these continents contains about a quarter of all Webpages. Often called the *central core*, it gives a home to all major Websites from Yahoo! to CNN.com. Its distinguishing feature is that it is easily navigable, since there is a path between any two documents belonging to it. This does not mean that there is a direct link between any two nodes of the central core. Rather, there is a path along nodes belonging to the core that allows you to surf between any two nodes.

The second and third continents, called *IN* and *OUT*, are just as large as the central core but much harder to navigate. From the pages of the *IN continent* you can reach the central core, but there are no paths from the core taking you back to IN land. In contrast, the nodes belonging to the *OUT continent* can be easily reached from the central core, but once you have left the core, there are no links to take you back. The OUT land is populated by corporate Websites that can be easily reached from outside; but once you get in, there is no way out. The fourth continent is made of *tendrils* and disconnected *islands*, isolated groups of interlinked pages that are unreachable from the central core and do not have links back to it. Some of these isolated groups can contain thousands of Web documents. About a quarter of all Web documents are located on such islands or tendrils. In general the location of a Webpage on the Web has little to do with the page's content; rather it is mostly determined by its relationship, via incoming and outgoing links, to other documents.

These four continents significantly limit the Web's navigability. How far we can get surfing depends on where we start. Taking off from a node belonging to the central core, we can reach all pages belonging to this major continent. No matter how many times we are willing to click, however, about half of the Web will still be invisible to us, since the IN land and the isolated islands cannot be reached from the core. If we step out of this core, into the OUT land, we will soon hit a dead end. If we start our journey from a tendril or an isolated island, the Web will appear very tiny because only the other documents on the same island will be reachable. If your Webpage is on an island, the search engines will never discover it, unless you submit your URL address to them.

Therefore, our ability to map out the full World Wide Web is not only a question of resources or economic incentives. The directedness of the links creates a very fragmented Web dominated by four major continents. Search engines have an easy time mapping out about half of it, the connected component and the OUT land, since the nodes belonging to them can be located starting from any node of the frequently visited central core. However, the other half of the Web, made up of the islands and IN land, is hopelessly isolated. No matter how hard the robots try, they will not be able to find the documents on them. This is why most search engines allow you to submit the address of your Website. If you do that, they can start crawling from it and potentially discover links to regions of the Web where they have never been. If you refuse to volunteer this information, many nodes could be residing in terra incognita for years to come.

Is this fragmented structure here to stay? Or will the evolving and growing Web eventually absorb the four continents into a single, fully connected core? The answer is simple: As long as the links remain directed, such homogenization will never occur. The continents are by no means a property peculiar to the World Wide Web. They appear in all directed networks. Consider for example a network crucial for our ability to find scientific information: the citation network. Each scientific paper cites other papers, relevant to the discussed work. A mathematics paper would cite other math papers focusing on similar problems or occasionally a biology or a physics paper, illustrating the applications of the obtained results. Therefore, all scientific publications are part of a web of science in which nodes are research publications connected by citations. These links are directed. Indeed, following the references at the end of this book will allow you to find the quoted papers. Yet none of these papers could send you to this book, since they do not cite it. The citation network is a very peculiar directed network in which the IN and OUT components reflect the historical ordering of the papers and the central component is very small if it exists at all. Nature also harbors some directed webs. In food webs, species are connected by links telling us which species feeds on which other species. The links of these networks seldom go both ways: The lion eats the antelope and never the other way around.

The bottom line is that *all* directed networks break into the same four continents. Their existence does not reflect any organizing principles particular to the Web. Random or scale-free, if the links are directed, the continents are there. This was recently demonstrated by Sergey Dorogovstev, José Mendes, and A. N. Samukhin, from the University of Porto, Portugal. They showed that the size and structure of these continents can be predicted analytically. Obviously, depending on the particular network's properties, the relative size of these continents varies. Yet, these results indicate that, no matter how complex and large the Web becomes, the continents are here to stay.

4.

In June 2000 Cass Sustein, a law professor at the University of Chicago, conducted a random survey of sixty political sites, finding that only 15 percent of them have links to sites with opposite views.

In contrast, as many as 60 percent have links to like-minded Webpages. A study focusing on democratic discourse on the Web arrived at a similar conclusion: Only about 15 percent of Webpages offer links to opposing viewpoints. Sustein fears that by limiting access to conflicting viewpoints, the emerging online universe encourages segregation and social fragmentation. Indeed, the mechanisms behind social and political isolation on the Web are self-reinforcing: They alter the Web's topology as well, segregating the online universe. Therefore, the four continents are not the only isolated structures of the Web. On a smaller scale, these continents are sprinkled with vibrant villages and metropoli. These are Websites brought together by a joint idea, hobby, or habitat, forming communities of shared interests. Jazz enthusiasts form a well-defined Web-based community, but so do bird-watchers. Religious fundamentalists in eastern Europe share virtual space with their ideological counterparts in the United States. Antiglobalization activists in Europe join forces with their peers in Japan to coordinate strategies and activities.

Communities are essential components of human social history. Granovetter's circles of friends, the elementary building blocks of communities, pointed to this fact. Lately, however, perhaps unrecognized by their members, such communities are increasingly recorded in the Web's topology. A side effect of our digital life is that our beliefs and affiliations are publicly available. Each time we link to a Webpage, we are endorsing its relevance to our intellectual curiosity. Thus the links of an enthusiastic bird-watcher can take us to other like-minded Web sites, allowing us to map out the community of bird enthusiasts.

Identifying such Web-based communities has tremendous potential for applications. Indeed, finding the community of sports car enthusiasts would allow car companies to most effectively market their new models by placing ads at several hubs of this community. AIDS activists could use community knowledge to mobilize those who passionately care about the disease, molding them into an effective lobbying and action group. Organizers of ethnic festivals could take advantage of information about Web-based ethnic communities to advertise upcoming events and incubate local grassroots organizations. The problem is that there are a billion-plus pages out there. Can we locate communities on such a gigantic web?

Supreme Court justice Potter Stewart famously remarked in 1964 that "I shall not today attempt further to define [obscenity] . . . and perhaps I never could succeed in intelligibly doing so. But I know it when I see it." We face similar problems when we try to find a proper definition of "Web-based communities." We all know them once we see them, but everybody has slightly different criteria for them. One reason is that there are no sharp boundaries between various communities. Indeed, the same Website can belong simultaneously to different groups. For example, a physicist's Webpage might mix links to physics, music, and mountain climbing, combining professional interests with hobbies. In which community should we place such a page? The size of communities also varies a lot. For example, while the community interested in "cryptography" is small and relatively easy to locate, the one consisting of devotees of "English literature" is much harder to identify and fragmented into many subcommunities ranging from Shakespeare enthusiasts to Kurt Vonnegut fans.

Recently Gary Flake, Steve Lawrence, and Lee Giles, from NEC, suggested that documents belong to the same community if they have more links to each other than to documents outside of the community. This definition is precise enough to develop algorithms to identify different groupings given the topology of the World Wide Web. It turns out, however, that actually finding these communities is notoriously difficult. This kind of search belongs to the class of so-called *NP complete* problems, which means that, though in principle communities can be located, there is no efficient algorithm for doing so. Therefore, the difficulty in finding communities on the Web is similar to solving the traveling salesperson's problem, which asks us to find the shortest route reaching a given number of cities assuming that we are not allowed to visit the same city twice. The only algorithm guaranteed to work for finding communities or the route for the traveling salesperson requires us to try all possible combinations. For communities, the time required to perform such a search increases exponentially with the size of the Web. With fast enough computers we might be able to locate communities in a sample of a hundred documents. Uncovering them from a billion Webpages, however, is simply out of the question.

Combining content and topology makes the problem somewhat less challenging. For example, we can focus on documents that contain only one or two keywords. Lada Adamic, from Stanford University, recently investigated communities discovered by searching for the phrases "abortion—pro choice" and "abortion—pro life." The pro-life query resulted in a core of forty-one documents in which you could get from each page to the other ones. In contrast, the pro-choice movement was fragmented into many disconnected sites.

Such differences in the structure of competing communities have important consequences for their ability to market and organize themselves for a common cause. As Adamic notes, a campaign against the partial birth abortion bill launched from the middle of the pro-life cluster could easily reach other pro-life sites, since there are many links between them. Furthermore, due to the links on the pro-choice sites, the visitors of pro-choice sites would also learn about it. However, one would need to advertise at several disconnected pro-choice Websites to achieve an equally efficient campaign against the bill. Therefore, not only does the pro-life community have a better presence on the Web, it is also better organized—its sites are more aware of each other.

Far from being a homogenous sea of nodes and links, the Web is fragmented into four continents, each of which hosts many villages and cities that appear as overlapping communities. Any of us willing to take up a virtual presence belongs to one or several of them. To be sure, we are far from fully understanding this fine structure of the Web. But many forces, from commercial interests to scientific curiosity, increasingly motivate us to do better. As we dig deeper, I am sure that we will encounter many surprises, offering us an even clearer view of this complex, amorphous, ever changing online universe.

5.

On November 20, 2000, in a precedent-setting decision, Judge Jean-Jacques Gomes of France ordered Yahoo! to deny French consumers access to any of its sites that auction Nazi memorabilia. It did so by upholding a French law prohibiting the sale of such items in France. The legal implications of the court's decision are still being debated across the world. Yahoo! argued that the Internet is fundamentally free from geographic and national boundaries and that subjecting the U.S. company to national laws around the world was therefore a severe breach of the Internet's basic philosophy. Others disagreed, saying that there is nothing particularly novel about the Internet and that it should be covered by the same international trade agreements as any international business.

Beyond the legal ramifications, the deeper issue is about the code—the software behind the Web. The French court acknowledged that considering the nature of the Web, there is no way to keep France completely isolated from the world. They were persuaded, however, by experts who testified that Yahoo! could put in place a filtering mechanism that would block at least 70 to 80 percent of French nationals trying to reach Yahoo!'s Nazi sites. Thus, the court ordered Yahoo! to alter the code. This is exactly the type of action that Lawrence Lessig, a Stanford University law professor, envisioned in his influential book *Code and Other Laws of Cyberspace*. According to Lessig, "Left to itself, cyberspace will become a perfect tool of control. . . . [T]he invisible hand of cyberspace is building an architecture that is quite the opposite of what it was at the cyberspace's birth."

Lessig uses the word *architecture* to mean the sum of all software running behind the Web, concluding that the only way to influence behavior in cyberspace is to regulate the code. He suggests that two forces are aligned to do just that. First, governments have a hard time policing behavior on the Web. It is easy to write legislation limiting access to everything from pornography to keys to cryptographic codes. In a borderless cyberworld, however, it is almost impossible to enforce these laws. If governments pass on the opportunity to regulate the Web, commerce will live with it. Companies seeking a more secure business environment in which they can identify customers for various purposes ranging from security concerns to marketing will push the code in the direction of control. Netizens will completely lose their anonymous and space-free existence as the technology develops to meet the merchants' desires.

On one hand, as the Yahoo! case and others have demonstrated, some of Lessig's bleak predictions have become reality. On the other hand, in my view, to truly understand cyberspace we need to distinguish carefully between *code* and *architecture*. Code—or software—is the bricks and mortar of cyberspace. The architecture is what we build, using the code as building blocks. The great architects of human history, from Michelangelo to Frank Lloyd Wright, demonstrated that, whereas raw materials are limited, the architectural possibilities are not. Code can curtail behavior, and it does influence the architecture. It does not uniquely determine it, however.

Like architects' buildings, the Web's architecture is the product of two equally important layers: *code* and *collective human actions* taking advantage of the code. The first can be regulated by courts, government, and companies alike. The second, however, cannot be shaped by any single user or institution, because the Web has no central design—it is self-organized. It evolves from the individual actions of millions of users. As a result, its architecture is much richer than the sum of its parts. Most of the Web's truly important features and emerging properties derive from its large-scale self-organized topology.

A good example is democracy on the Web. We've seen that the scale-free topology means that the vast majority of documents are hardly visible, since a highly popular minority has all the links. Yes, we do have free speech on the Web. Chances are, however, that our voices are too weak to be heard. Pages with only a few incoming links are impossible to find by casual browsing. Instead, over and over we are steered toward the hubs. It is tempting to believe that robots can avoid this popularity-driven trap. They could, but they don't. Instead, the likelihood that a document will be indexed by a search engine depends strongly on the number of its incoming links. Documents with only one incoming link have less than a 10 percent chance of being noticed by any search engine. In contrast, robots find and index close to 90 percent of pages that have twenty-one to one hundred incoming links.

Lessig is right: The architecture of the Web controls just about everything, from access to consumers to the probability of being visited by surfing along the links. But the science of the Web increasingly proves that this architecture represents a higher level of organization than the code. Your ability to find my Webpage is determined by one factor only: its position on the Web. If many people find my page interesting and they link to me, my node will slowly turn into a minor hub, and search engines will inevitably notice. If everybody ignores my Website, so will the search engines. I will join the ranks of invisible Websites, which are the majority anyway. Thus the Web's large-scale topology—that is, its true architecture—enforces more severe limitations on our behavior and visibility on the Web than government or industry could ever achieve by tinkering with the code. Regulations come and go, but the topology and the fundamental natural laws governing it are time invariant. As long as we continue to delegate to the individual the choice of where to link, we will not be able to significantly alter the Web's large-scale topology, and we will have to live with the consequences.

6.

The great thing about the Web is that our Webpages mature with us. Once we alter our personal page, nobody can haunt us with the opposite views we might have held decades earlier. Do you remember that boyfriend you broke up with a few years ago? Of course you do, but you probably hope that nobody else does. To be sure, all his pictures are gone from your Webpage. How about that high-school manifesto you are still embarrassed about? Or that collection of links to Democratic sites you assembled a mere two years before running on a Republican ticket? They are all untraceable. Or at least, we tend to think so. That is because most netizens have never heard of Brewster Kahle. The truth is, Kahle could easily have a copy of all the pictures and documents that you so carefully removed from your Website and have now forgotten.

The inventor of wide area information servers and founder of Alexa Internet, one of the major search engines, Kahle is a veteran of the Web. After selling Alexa to Amazon.com in 1999, he used the proceeds to create the Internet Archives, a nonprofit organization located in Presidio, a converted military base in downtown San Francisco. His goal is simple: He wants to prevent the Web's content from disappearing into the past.

When I visited the Archives to give a talk at the First Internet Archive Workshop in March 2000, Kahle reminded me of the ancient library of Alexandria. It was believed to have had a copy of all books written in the ancient word, all of which disappeared when the library was burned to the ground. He also told me about great cinematographic collections that were recycled for their silver content. Without cultural artifacts, humanity has no memory, and without memory it cannot learn from its successes and failures. When it comes to the World Wide Web, we are again letting history go unrecorded. To avoid repeating history, Kahle's brainchild, the Internet Archives, carefully keeps all documents that Alexa has crawled to since 1996. The collection has already swelled to 100 billion Webpages, representing about 100 terabytes of information. In comparison, all books and documents archived by the Library of Congress are only about 20 terabytes.

The Archives' collection is of unparalleled value for historians, social scientists, and Web topographers alike. To write the history of the 2000 presidential election, you would start with the Archives. They have a time machine that allows you to see the candidate's sites, voter guides, and the Web pages of political parties, exactly as they were during the campaign. Do you want to track the reaction of the online universe to the September 11, 2001, terrorist attacks? One month after the events the Archives already had a collection of 200 million related documents. If you are a Web topographer aiming to understand the Web's architecture, the Archives are an excellent starting point. They let you trace when and where Webpages and links were added and removed, how some latecomer nodes become popular overnight, and how former hubs lost their shine. Comparing the maps of the Web taken at different time intervals, you can follow the emergence and crystallization of virtual communities. The Archives have the data to reconstruct the chaotic evolution of nodes and links, helping to uncover the mechanisms responsible for the Web's current architecture.

The Archives have many fans from many different disciplines, but most researchers who could take advantage of them either do not know of their existence or lack the programming skills to access and efficiently use them. So their full potential is still untapped by researchers and the public alike. However, I hope that the Internet Archives are only the beginning of an awakening to our historical responsibility towards the online universe. The Archives are far from capturing everything out there. Their main collection comes from Alexa, the search engine founded by Kahle and Bruce Gilliat in 1996. As we already know, search engines cover only a small fraction of the World Wide Web, and Alexa was never known for pursuing a significant coverage. Therefore, despite their enormous size, the Archives' current collection represents only a tiny fraction of the Web, mostly popular Webpages. Alexa got the hubs; the rest, the vast majority of less connected pages ignored by their robots, are slipping into oblivion at a rate of millions per day.

7.

To an alien approaching our solar system, the Earth would appear to be nothing more than a spherical ball. Getting closer, the alien might start noticing the continents. The bright lights of Paris, New York City, London, and Tokyo offer clues of intelligent life. Getting even closer, smaller communities become discernable, and a fine structure of connecting highways and roads emerges. The alien would have to come really close, however, to see the human beings responsible for the large-scale order visible from space.

Our exploration of the World Wide Web has followed an identical route. First we discovered the inhomogeneous large-scale topology and understood that it is as unavoidable as the spherical shape of most planets. Looking closer, we noticed four major continents, each obeying different laws. Bringing more details into focus, we started to see communities, groups of Webpages held together by common interests. These forays into the unknown have significantly altered our understanding of the World Wide Web. We learned that the online universe is much larger than anyone ever anticipated. It also grows faster than we were ready to believe. To our dismay, we also found that it is much less charted than we were willing to accept. Two years ago, six out of ten pages had not been visited by any search engine. If the trend can be trusted, today's search engines see an even smaller fraction of the Web. The good news is that competition forces the search engines to do a better job. But we should never lose sight of the big picture: Whatever the extent of their competition, the Web is even bigger.

Yet we shouldn't underestimate the enormous services the search engines and their robots offer us. We often sigh in desperation, calling the Web a "jungle." The truth is, without robots it would be a black hole. Space would curve around it such that anything falling in would never get out. Robots keep the World Wide Web from collapsing under its increasing complexity. They fold the space out, maintaining order in the chaos of nodes and links.

Our life is increasingly dominated by the Web. Yet we devote remarkably little attention and resources to understanding it. Relatively little effort would be required to bring along a new revolution in information access. It will happen. The question is, what do we lose in the meantime?

In an increasingly Internet-dominated society, understanding the World Wide Web has tremendous value in and of itself. For me, however, the rewards go beyond that. One of the most exciting aspects of this exploration has been uncovering laws whose validity does not stop at the gates of cyberspace. These laws, applying equally well to the cell and the ecosystem, demonstrate how unavoidable nature's laws are and how deeply self-organization shapes the world around us. By virtue of its digital nature and enormous size the World Wide Web offers a model system whose every detail can be uncovered. We have never gotten this close to any network before. It will continue to be a source of inspiration and ideas to anybody aiming to grasp the properties of our weblike universe.

THE THIRTEENTH LINK

The Map of Life

IN FEBRUARY 1987 the journal *Nature* reported a landmark discovery: the gene for manic depression, or by its more recent name, bipolar disorder. Manic depression affects 1 to 5 percent of adults in the United States, and as many as 25 to 50 percent of those attempt suicide at least once. Because the risk of developing manic depression is five to ten times higher if first-degree relatives have the disease, the prevailing view is that manic depression is a genetic disorder. So as soon as methods for linking illnesses to specific genes emerged, the race was on to find the manic depression gene. The much coveted "first" seemed to have gone to the authors of the 1987 *Nature* paper, who located the gene on chromosome 11 while studying a large Amish family in Lancaster, Pennsylvania. Yet two years later the research group recanted the results. The blunder did not discourage other gene hunters, however. If anything, it gave them extra motivation to find the real gene. In 1996, almost a decade after the first published study, three independent research groups reported links to genes on other chromosomes. Another Amish study implicated chromosomes 6, 13, and 15; a study focusing on the isolated population of Costa Rica's Central Valley documented links to chromosome 18; and results derived from a large Scottish family indicated the involvement of chromosome 4. Research on another prominent mental disorder, schizophrenia, followed a similar pattern, linking the disease to two different regions of chromosome 1, with a different research group implicating chromosome 5 a few years later.

Absentminded scientists? Bad research? Far from it. These are not conflicting results. They simply demonstrate that most illnesses, ranging from manic depression to cancer, are not caused by a single malfunctioning gene. Rather, several genes interacting through a complex network hidden within our cells are simultaneously responsible. Faced with the gigantic task of figuring out the building blocks of the cell, from genes to proteins, scientists until recently focused on biology rather than networks. But with the pieces now in hand, postgenomic biology is taking a step back to grasp the big picture. New and exciting discoveries that are revolutionizing biology and medicine tell us loud and clear: If we want to understand life—and ultimately cure disease—we must think networks.

1.

"Today we are learning the language in which God created life," said President Bill Clinton on June 26, 2000, at the White House ceremony announcing the decoding of the 3 billion chemical "letters" of the human genome. Is it true? Has humanity been handed the "book of life"? Are Francis Collins and Craig Venter, the two gentlemen who stood on either side of the president, the prophets of the twenty-first century? After all, Collins and Venter, representing the publicly funded Human Genome Project and the private Celera Genomics, which each decoded the human genome, brought the book to us.

Open the "book of life" and you will see a "text" of about 3 billion letters, filling about 10,000 copies of the *New York Times* Sunday edition. Each line looks something like this:

TCTAGAAACA ATTGCCATTG TTTCTTCTCA TTTTCTTTTC ACGGGCAGCC

These letters, abbreviations of the molecules making up the DNA, could easily mean that the anonymous donor whose genome has been sequenced will be bald by the age of fifty. Or they could reveal that he will develop Alzheimer's disease by seventy. We are repeatedly told that everything from our personality to future medical history is encoded in this book. Can you read it? I doubt it. Let me share a secret with you: Neither can biologists or doctors.

To be sure, the sequencing of the human genome is a triumph, the result of modern molecular biology's ability to reduce complex living systems to their smallest parts. It is undoubtedly a catalyst of a new era in both medicine and biology. But the genome project has brought along a new realization: The behavior of living systems can seldom be reduced to their molecular components.

Our inability to find a single gene responsible for manic depression is the best illustration. A list of suspected genes is not sufficient. To cure most illnesses, we need to understand living systems in their integrity. We need to decipher how and when different genes work together, how messages travel within the cell, which reactions are taking place or not in any given moment, and how the effects of a reaction spread along this complex cellular network. To achieve this we must map out the network within the cell. This web of life determines whether a cell develops into skin or labors constantly in the heart, decides the cell's response to external disturbances, holds the key to survival in constantly changing environments, tells the cell when to divide or die, and is responsible for illnesses ranging from cancer to psychiatric disorders. As the historic *Science* article that reported the decoding of the human genome concluded, "there are no 'good' genes or 'bad' genes, but only networks that exist at various levels."

2.

The decoding of the human genome offered us an inventory of the cell's parts. To return to our car analogy, it is like having thousands of car parts in your backyard. If you ever want to see that car running again, you must find the blueprint, a map telling you how to assemble it. For most cells this map is almost as elusive now as it was fifteen years ago at the beginning of the Human Genome Project. The absence of a cellular search engine is only part of the problem. The biggest difficulty is that within each cell there are many layers of organization that can each be viewed as a complex network. To understand the web of life, we need to acquaint ourselves with some of these.

In today's weight-conscious society, it is common knowledge that cells burn food by splitting complex molecules to create the cells' building blocks and the energy they require to stay alive. This is achieved through a web of hundreds of multistep intracellular biochemical reactions, together referred to as the *metabolic network*. The nodes of this network can be simple chemicals, such as water or carbon dioxide, or more complex molecules made of dozens of atoms, such as ATP. The links are the biochemical reactions that take place between these molecules. If two molecules, A and B, react with each other to create C and D, then all four of them are connected in the cell's complex metabolism.

Think of the cellular metabolism as the engine in your car. Having an engine in and of itself will not get you very far. You need wheels, suspension, brakes, lights, and many other components, each ensuring that the car will run safely on the road. In a similar vein, the cell has an intricate *regulatory network* that controls everything from metabolism to cell death. The nodes of this network are the genes and the proteins encoded by the gigantic DNA molecule. The links are the various biochemical interactions between these components. The genes are first copied into unique messenger RNA molecules, which are then translated into proteins. Some proteins interact with the DNA, initiating or suppressing the translation of new genes, repairing accidental DNA damage, copying the two strands of DNA when the cell replicates, and so on. Other proteins interact with each other, forming large protein complexes. A prominent example is hemoglobin, a protein complex made of four proteins that bind together to transport oxygen in our bloodstream. Therefore, proteins can be viewed as nodes of a complex protein-protein interaction network in which two proteins are connected if they can physically attach to each other. The full weblike molecular architecture of a cell is encoded in the *cellular network*, a sum of all cellular components (genes, proteins, and other molecules), connected by all physiologically relevant interactions, ranging from biochemical reactions to physical links. This web of life contains all metabolic, protein-protein, and protein-DNA interactions present in the cell.

Not too long ago it was widely believed that everything that matters for an organism's biological history is encoded in the genes. Postgenomic biology, though still in its infancy, is already fighting an important battle. It aims to diminish the all-encompassing role historically attributed to individual genes. Genes are known to play a *structural* role, determining the scope and make of proteins and passing this information in a hereditary manner to subsequent generations. Recently, however, scientists have discovered that genes also play an important *functional* role as members of a complex cellular network. This functional role is apparent only in the dynamic context in which an individual gene interacts with many other cellular components. The gene's structural role can be unearthed from its sequence. We now have the complete sequence for several key organisms, ranging from *Escherichia coli* bacteria to humans. We are only at the beginning, however, of the second, equally revolutionary scientific endeavor: uncovering the gene's functional role. To achieve this we need a second genome project, this time mapping the web within the cell. We have the "book of life." Now we need the *map of life*.

3.

Zoltán Oltvai, a cell biologist at Northwestern University Medical School in Chicago, had several significant and much cited discoveries under his belt when we met in 1998. At that time we both lived in Oak Park, a Chicago suburb styled by the towering architectural presence of Frank Lloyd Wright. With small children of similar age, we started to visit each other regularly. After exhausting all topics related to culture and politics, our conversation turned to science and biology. By then my group was pressing ahead with research on the Web and Internet. Inevitably, our weekend chats drifted toward the similarities and differences between the web of life and other complex networks. Soon an ongoing argument developed. The Web and the actor network are scale-free because they emerged thanks to growth and preferential attachment, processes that are easily identifiable in both networks. The cell, on the other hand, is different. To be sure, the original assembly of the first protocells from a primordial soup of organic molecules might have resembled a growing network. But during the past three billion years evolution and natural selection took their course. During this time there was significantly less growth, just a lot of tinkering with the cellular network, streamlining and optimizing it. Thus, on the one hand, even if a scale-free topology had developed when lifeless molecules took their first steps towards life, it might have been lost because of the all-encompassing effects of evolution. On the other hand, it is hard to fathom that the complex biochemical web within the cell would be completely random. So is the map of life, like the Erdős-Rényi network, random, or is it scale-free, like the Web? How do we characterize the cell's complex topology?

After we ran out of arguments to convince ourselves one way or another, Oltvai and I decided to move our discussions off the playground and look for real data on the web of life. Fortunately, for most of the twentieth century, biology and biochemistry were devoted to identifying and interrelating the various molecules within the cell. James Watson, the codiscoverer of the double helix structure of DNA, wrote in 1970 in the now classic *Molecular Biology*, "We already know at least one-fifth, and maybe more than one-third of all metabolic reactions that will ever be described [in *E. coli* bacteria]," suggesting that "within the next ten to twenty years we shall approach a state in which it will be possible to describe essentially all metabolic reactions." Watson's vision has been fulfilled. Today bacteriologists believe that the complex network of more than seven hundred nodes and close to a thousand links represents pretty much the full list of reactions fueling the *E. coli* metabolism. What Watson could not have imagined in 1970 is that thirty years later online databases would be compiling the network of metabolic reactions for hundreds of organisms. While we are still missing a detailed metabolic map of the highly complex human cell, our knowledge of several simpler organisms is close to being complete.

So my discussions with Oltvai could not have been better timed. A few years earlier my lab's quest to study the cell's topology would have been brought to a halt by an absence of data. In late 1999, however, several Websites had the maps we were looking for. After researching the available databases, we settled on a new one, run by the Argonne National Laboratory outside Chicago, nicknamed "What Is There?" which compiled the metabolic network of forty-three diverse organisms. Hawoong Jeong, once again displaying his computer wizardry, wrote a program that downloaded each reaction individually. Oltvai and I watched over his shoulder as he made sense of this extremely complex web, assembling one by one the full metabolic map for these forty-three organisms. Having finished that, he moved on to characterize these networks, calculating how many reactions each molecule participates in. The robustness of the results was shocking. No matter which organism we examined, a clear scale-free topology greeted us. Each cell looked like a tiny web, extremely uneven, with a few molecules involved in the majority of reactions—the hubs of the metabolism—while most molecules participated in only one or two.

4.

To harken back to our social networks, if two molecules participate in the same reaction, their separation is one. If, however, two subsequent reactions are needed to connect them, their separation is two. Putting all nodes and links together, will this complex network within the cell have small-world properties?

Measuring the separation between molecules is not an outgrowth of our obsession with six degrees of separation. The diameter of the network—or degree of separation between nodes—has biological significance. For instance, if we should find that the shortest chemical path between two molecules is one hundred, then any change in the concentration of the first molecule will have to go through one hundred intermediate reactions before reaching the second molecule. Any perturbation will decay and die along such a long path.

To our great surprise, the measurements indicated that the typical path lengths are much shorter than one hundred. In fact, cells are small worlds with *three degrees of separation.* That is, most pairs of molecules can be linked by a path of three reactions. Perturbations, therefore, are never localized: Any change in the concentration of a molecule will shortly reach most other molecules. This finding was supported by the study of Andreas Wagner, from the University of New Mexico, and by David A. Fell, from Oxford Brooks University, who independently concluded that the *E. coli* metabolic network is scale-free and has small-world properties.

Though unexpectedly short, the *three* degrees was not the most interesting aspect of our finding. Because the forty-three organisms all had different sizes, we expected that the separation would increase with the organism's size, just as the Web's diameter increases with the number of documents. Surprisingly, the measurements indicated that whether we are navigating the tiny network of a small parasite bacterium or the highly developed highway system of a multicellular organism, such as a flower, the separation is the same. Although the difference in the cellular architecture between a primitive bacterium and a cell from a multicellular organism could be as large as the difference between a tiny village and New York City, stripped to their dynamically relevant networks, all cells feel like a small town. Digging deeper, we learned that most cells share the same hubs as well. That is, for the vast majority of organisms the ten most-connected molecules are the same. Adenosine triphosphate (ATP) is almost always the biggest hub, followed closely by adenosine diphosphate (ADP) and water.

To be sure, the role of ATP, ADP, and water as prominent hubs was by no means surprising. In cells, ATP serves as a convenient and versatile store of energy, driving hundreds of biochemical reactions. By supplying energy to these reactions, ATP turns into ADP by giving up a phosphate group; thus, within the metabolic web, both ATP and ADP are linked to a huge number of molecules participating in energy-hungry reactions. Yet, taken together, the top-ten list of highly connected molecules was rather revealing. A key prediction of the scale-free model is that nodes with a large number of links are those that have been added early to the network. In terms of metabolism this would imply that the most connected molecules should be the oldest ones within the cell. And indeed, the analysis of Wagner and Fell has shown that the most-connected molecules have an early evolutionary history as well. Some of these molecules are believed to be the remnants of the so-called RNA world, the evolutionary step before the emergence of DNA, while others are known to be the components of the most ancient metabolic pathways. Therefore, the first mover advantage seems to pervade the emergence of life as well.

If all organisms have the same scale-free topology and the same node separation and share the same hubs, how do cells of different organisms differ from one another? Is there any difference between the chemical architecture of a bacterium and that of a human cell? It turns out that there are significant differences. Comparing the metabolic network of all forty-three organisms, we found that only 4 percent of the molecules appear in all of them. Though the hubs are identical, when it comes to the less connected molecules, all organisms have their own distinct varieties. Life looks like a suburb in which each house was designed by the same architect, but different builders and interior designers were commissioned to offer the finishing touches, from the material of the floor to the size and make of the windows. In an aerial photograph all houses appear to be alike. The closer you get to them, however, the more you start noticing the differences.

Metabolism represents only one component, albeit an important one, of the cellular network. Will the same scale-free architecture also be present in the regulatory network—the web responsible for running the cell? Indeed, we are ultimately interested in the full weblike molecular architecture of living organisms. The question is, do the different components of this web of life follow the same laws and architectural features, or has evolution discovered different solutions for the various components? Beyond our desire to comprehend the fundamental features of the cell's architecture, understanding the regulatory network has important practical implications as well. Indeed, genetic disorders result from malfunctions of the nodes of the regulatory network. Therefore, the robustness of this network to node failures determines our ability to survive various diseases, as well as researchers' ability to design drugs that can cure those disorders that we cannot easily tolerate.

5.

Baker's yeast, one of the simplest eukaryotic cells, has about 6,300 genes, encoding about the same number of proteins. Though this is only a fifth of the estimated 30,000 different genes a human cell contains, it is already an enormous number. In general, when proteins interact by sticking to one another, they have a good reason for doing so. Most interactions play some important functional role in the cell's life. Therefore, to understand how cells work we must identify all pairs of proteins that can interact. For baker's yeast, that requires checking 6,300 times 6,300 pairs—close to forty million potential interactions. With standard molecular biology tools this would take decades and hundreds of people. Yet, despite the magnitude of the job, two research groups have independently obtained a detailed map of the yeast protein network. They succeeded thanks to an important technological breakthrough, the so-called two-hybrid method. Developed by Stanley Fields in 1989, the two-hybrid method offers a relatively rapid semiautomated technique for detecting protein-protein interactions. Though the method is known to provide numerous false negatives and positives, the map it generated offers an unprecedented opportunity to peek into the cell's regulatory organization.

Electrified by the insights offered by the topological analysis of cellular metabolism, in the fall of 2000, Oltvai, Jeong, and I, together with a young student, Sean Mason, became interested in the structure of the protein interaction network. The two-hybrid data, published a few months earlier, offered an excellent opportunity for such a study. After downloading all known protein-protein interactions, we reconstructed the protein network of yeast with the aim of studying its large-scale features. Once again, the results left little room for ambiguity: They demonstrated that the protein interaction network has a scale-free topology. That is, most proteins in the cell play a very specific role, interacting with only one or two other proteins. A few proteins, however, are able to physically attach to a huge number of other proteins. These hubs are crucial for the cell's proper functioning and survival. Indeed, we were able to show that removing a gene responsible for a hub protein kills the cell 60 to 70 percent of the time. Mutations affecting a weakly connected protein, in contrast, have a less than 20 percent likelihood of proving lethal.

A series of parallel results supported these findings. Andreas Wagner independently confirmed that the yeast protein network has a scale-free topology. Stefan Wuchty, a young researcher working at the European Media Laboratories, found a similar architecture in a markedly different network within the cell. In his so-called protein domain network, the nodes are different facets through which proteins link to each other, two facets being considered connected if they are simultaneously present on the same protein. Jong Park and collaborators from the European Bioinformatics Institute in the United Kingdom spotted a scale-free topology when they reconstructed the yeast network from protein interaction data collected by the Protein Data Bank. Our research group has found the same structure in an organism very different from yeast, a simple bacterium called *Helicobacter pylori*, suggesting that the scale-free nature of the protein interaction network is a generic feature of all organisms.

Taken together, the similar large-scale topology of the metabolic and the protein interaction networks indicate the existence of a high degree of harmony in the cell's architecture: Whichever organizational level we examine, a scale-free topology greets us. These journeys within the cell indicate that Hollywood and the Web have only rediscovered the topology that life had already developed 3 billion years earlier. Cells are really small worlds that share the topology of many other nonbiological networks, as if the architect of life could design only these.

How did life arrive at this architecture? Almost as soon as we asked the question, we had an answer. Approximately a half year after the publication of our findings on the topology of the protein interaction network, I received three e-mails within about a month. Each of them contained a manuscript by a different research group. Amazingly, each of the three research groups independently offered the same simple and elegant explanation, claiming that the cell's scale-free topology is a result of a common mistake cells make while reproducing.

6.

Cells reproduce by duplicating their content and dividing into two. The details of these processes may vary for simple bacteria and more complex human cells. Certain steps are universal, however. First, in order to produce a genetically identical offspring cell, the DNA must be faithfully replicated. This process is not free of errors, however. Although the cell's intricate copying mechanism insures that DNA sequences are inherited with extraordinary fidelity, about one letter in a thousand is randomly changed every 200,000 years. Another common error is gene duplication. Through a rare accident in the copying process, gene duplication can occur when the ends of broken DNA molecules join together. As a result, segments of varying length of the parent DNA will appear twice in the offspring's genome. Such copying mistakes sometimes kill the cell. In other cases, multiple copies of the same gene have evolutionary advantages and are passed on to future generations. Hemoglobin is a well-known example.

Originally cells had only one hemoglobin gene. About 500 million years ago, during the evolution of the higher fish species, a series of gene duplications occurred, resulting in four copies of the hemoglobin gene scattered along the genome. Today each of these genes encodes one of the four components of the hemoglobin protein complex.

Gene duplication has a significant impact on the cellular network. It results in two identical genes, which produce identical proteins, that in turn interact with the same proteins. A new node thus has been created, the protein generated by the duplicated gene. Its neighbors, the proteins with which the duplicated protein interacts, will each now interact with both the parent and the identical offspring protein. Therefore, each protein in contact with the duplicated protein gains an extra link. In this game highly connected proteins have a natural advantage: They are more likely to have a link to the duplicating protein than their weakly connected cousins. It's not that hubs duplicate more often. Rather, since the hubs are in contact with more proteins, they are more likely to have a link to a duplicating node, which offers them an extra link, a subtle version of preferential attachment.

The most important feature of this explanation is that it traces the origin of the scale-free topology back to a well-known biological mechanism, gene duplication. It does so by showing that gene duplication can simultaneously lead to both the growth of the protein network by adding an extra protein and to preferential attachment by adding new links at a higher rate to the more connected proteins. It is too early to determine if this is the only explanation, since it is conceivable that different mechanisms, yet unexplored, could generate the same topology. It is unclear if it explains the scale-free structure seen in the metabolism, as well. Nevertheless, it demonstrates that mechanisms present in the cell can generate the scale-free topology. Therefore, at this point we are ready to turn to the next important question: Will the map of life help us better understand diseases and enhance our ability to eventually cure them?

7.

Cancer is the most researched human illness ever. The extraordinary attention the medical community has devoted to it has resulted in several significant breakthroughs. Probably the most important is the discovery of the p53 gene. Though reported as early as 1979 by David Lane and Arnold J. Levine, it was not until the late 1980s following the work of Bert Vogelstein that its role in cancer was fully appreciated. Vogelstein recognized that the p53 protein, created by the p53 gene, is a tumor suppressor. Just as your brakes allow you to stop your car, tumor suppressor genes act to slow and halt DNA replication and division into new cells. Healthy cells keep a small number of p53 molecules around. If radiation or some other injury damages the cell, more p53 is produced, preventing the progression of the cell through cell division. This gives the cell time to repair the damage before further copies of the malfunctioning cell can be produced. However, if the damage is irreparable, the p53 protein will activate a group of genes to kill the cell.

If the cell's brake—the p53 protein—malfunctions, the cell can run amok. Cancerous cells differ from healthy cells in their ability to multiply at a very high rate. Indeed, about 50 percent of human cancers contain mutations in the p53 gene. This observation has stimulated an avalanche of research, resulting in over 17,000 publications since 1989. In recognition of its central role in cancer, in 1993 the p53 molecule was named "Molecule of the Year" by *Science*. Considering the attention the p53 molecule has received, one might have expected that a cure for cancer would have been found by now. After all, all we need to do is to develop drugs that make sure the p53 molecule always does its job. Why, then, has this huge amount of research not yet translated into a universal cancer drug?

Despite its important role in human cancer, fixing the p53 gene alone will not lead to a cure for this deadly disease. The reason was recently articulated by the very people responsible for placing p53 at the center of cancer research. Vogelstein, Lane, and Levine in November 2000 coauthored a *Nature* paper that made networks the crux of their argument. The reason why we do not fully understand cancer, the three suggested, is that the cell is like the Internet.

The three researchers argued that we must stop our obsession with the omnipresent p53 molecule and focus instead on what they called the *p53 network*, a sum of all molecules and genes interacting with the p53 molecule. As they put it, "One way to understand the p53 network is to compare it to the Internet. The cell, like the Internet, appears to be a 'scale-free network': a small subset of proteins is highly connected (linked) and controls the activity of a large number of other proteins, whereas most proteins interact with only a few others. The proteins in this network serve as the nodes, and the most highly connected nodes are the hubs. In such a network, performance is almost unchanged by random removal of nodes. But such systems contain an Achilles' heel."

The "Achilles' heel" of a network, you'll recall, refers to the vulnerability of its hubs. The inactivation of less connected molecules does not have draconian effects on the cell, whereas a mutation in the p53 molecule, one of the clear hubs of the cellular network, turns the cell cancerous and eventually kills the organism. This explains why combined pharmaceutical attacks on molecules that interact with the p53 molecule have progressively more severe effects on the cell, resembling an attack on the p53 molecule itself.

Vogelstein, Lane, and Levine's *Nature* paper demonstrated the strength and ubiquity of network thinking. Ideas developed to better defend the Internet and quantify the effects of hacker attacks have fallen on fertile ground in cell biology, which is concerned with the defense of healthy human cells against all threatening organisms. At the heart of Internet research and cell biology, the questions are similar. The first step is to map out the network behind these systems. Then from these maps we need to infer the laws that govern the networks. At that point the Internet topographer, the Web mapper, and the cancer researcher will be in the same camp.

Yet the most important implication of the p53 network goes beyond the fundamental analogies it illuminates between cells and the Internet. It points to a new approach to drug therapies and drug development. The ultimate goal of studying the p53 network is to find a cure for cancer. As we discuss next, this is largely a trial-and-error process. In most cases cancer therapies aim for destruction: They kill the cancerous cells by disrupting their cellular network with either drugs or radiation. The increasing understanding of the p53 network suggests another avenue: We must first decipher the precise topology of this network, fully understanding all interactions. With such a map in hand, we can start a frontal attack, finding drugs that restore the functions of the p53 molecule without dismantling the network around it.

8.

Until recently we could only treat the symptoms of illnesses like cancer, heart disease, and psychiatric disorders. We searched for rare chemicals everywhere, from chemistry labs to rain forests, hoping that they would offer miracle drugs for some diseases. According to some estimates, the drugs available on the market target only about 500 of the 30,000 proteins in the human body. And though we have multiple drugs for many diseases, it is often a trial-and-error process to figure out which works for a given patient.

A detailed understanding of the full biochemical network within the cell promises to eliminate this guesswork. With knowledge of the precise wiring diagram of a cell and diagnostic tools capable of capturing the strength of the various cellular interactions, doctors in the future could test the response of your cells to a drug before you even take it. Thanks to the map of life, which implies a detailed understanding of how genes work together, we will someday be able to diagnose diseases like manic depression or cancer before any of the symptoms have occurred. This knowledge will help us develop drugs that are so fine-tuned and highly precise that they affect only the malfunctioning cells, leaving the healthy cells alone. In other words, they will provide real cures.

Changing the concentration of a chemical in your body via a drug could reduce the symptoms of a particular disease. However, since the cell is controlled by a complex network with small-world properties, a drug-induced perturbation inevitably affects many other chemicals, possibly creating undesired side effects. Patients treated for manic depression might die of heart disease, a condition they had never experienced before. Furthermore, the drug that causes heart disease for you could have no side effects on another individual. We all have different eye and hair colors and facial features, after all, so it is not surprising that we metabolize drugs differently as well. With the map of life in hand and with tools such as the recently developed DNA chips that monitor the links between the genes, doctors will be able to obtain a detailed list of all molecules and genes affected by a given drug. Exploring side effects will no longer be guesswork. We will have personalized medicine, allowing the marketing and approval of drugs that are effective for only 10 percent of the population and potentially lethal for everybody else.

9.

If you suffered from manic depression in recent years, your first visit to the doctor probably started with an hour-long discussion to carefully examine your thoughts and feelings. Eventually you walked away with a drug prescription. If you had never appreciated how much of your brain's activity and well-being was a matter of brain chemistry, now, after taking the drug, you did. A layer of chemicals injected into your body rapidly took over your behavior and impulses. You discovered yourself doing things and having feelings that you never experienced before. In most cases the first drug didn't work. It perhaps made you hyperactive or even more depressed. A few weeks later the drug had to be switched for another, in hope of better results. Patients would routinely try five or six drugs over a period of several months before finding the one that worked best. While they made you feel better, these drugs didn't cure your illness. They temporarily altered your brain's chemistry, offsetting the changes caused by the malfunctioning of your genetic network. If you stopped taking them, the chemical imbalance would return, along with the symptoms of manic depression.

Twenty years from now things could look quite different. Facing the same doctor, you will have a five-minute discussion, just as you do in cases of simple influenza. An assistant will take a few drops of blood, and you will walk home empty-handed. In the evening you will pick up the medicine from the nearest pharmacy. The next day you will wake up fresh and happy, as you did before your symptoms appeared. Both the manic and the depressive you will have been washed away.

How will this breakthrough come about? First, the full biochemical network of the human cell will have been mapped by then, allowing us to understand in detail how different genes and molecules work together. Second, DNA and protein chips, new technologies now under development, will be in each doctor's office, allowing her or him to monitor which genes and proteins malfunction in your cells. While mapping the human cellular network will probably take over a decade, the instant monitoring of gene activity is already possible in some research labs.

By 2020 these advances will change medicine across the board. Kids will not be taken to the doctor with a sore throat—Mom will have a handheld device, with a replaceable chip, that will reveal that Tommy's sore throat is a streptococci infection, identifying the strain as well. She will be able to link the device to the computer and e-mail the profile to the doctor's office, so when Tommy shows up for school the drug is ready for him in the nurse's office. Most important, Tommy's drug will not be a strafing antibiotic that kills all bacteria, harmful or not, in his body. It will be designed and mixed on the spot to take out only the organism that made Tommy's throat hurt. It will be ineffective against any other bacteria, minimizing the chance that Tommy will develop an antibiotic resistance.

I don't believe that this vision is far-fetched. In fact, it is rather modest, perhaps even shortsighted. It is only a simple interpolation of the tools already present in most research laboratories around the world. These advances are rooted in a fundamental shift in how we look at everything from life to disease. They are the result of seeing the cell as a whole—as a network—rather than a bag of independent chemicals.

10.

The genome project is the ultimate celebration of the *gene*. Until recently we believed that the complete biological history of a human being was encoded in the 3 billion letters of the helical DNA. To be sure, the mapping of the human genome revolutionized biological research. But it also showed us what a small fraction of the vast world is really known to us and how much more is left to be explored.

In 1996 the decoding of the yeast genome gave the scientific community a shock: It contained as many as 6,300 genes. Only about a quarter of these were expected and could be assigned vague functions. To be on the safe side, and boosted by humans' perceived importance as the pinnacle of evolution, biologists estimated that the human genome would have at least 100,000 genes. This number was believed to be sufficient to account for the high complexity of *Homo sapiens*. Then came February 2001 and the publication of the human genome. It turned out that we have less than a third of the anticipated genes—only about 30,000. Therefore, a mere one-third increase in genes must explain the difference between us and the unsophisticated *Caenorhabditis elegans* worm—quite a provocative idea when we consider that the 20,000 genes of *C. elegans* need to encode only three hundred neurons, whereas our extra 10,000 genes have to account for the billion nerve cells present in our brain.

In short, it is now clear that the number of genes is not proportional to our perceived complexity. Then what does complexity mean? Networks point to the answer. Framed in terms of networks, our question becomes: How many different potentially distinct behaviors can a genetic network display with the same number of genes? In principle, two cells that are identical except that a specific gene is *on* in the first cell and *off* in the second could behave differently. Assuming that each gene can be turned on or off independently, a cell with *N* genes could display 2^N distinct states. If we adopt as a measure of complexity the potential number of distinct behaviors displayed by a typical cell, the difference between the worm and humans is staggering: Humans could be viewed as $10^{3,000}$ times more complex than our wormy relatives!

Whereas the twentieth century was seen as the century of physics, the twenty-first is often predicted to be the century of biology. A decade ago it would have been tempting to call it the century of the gene. Few people would dare say that any longer about the century we have just entered. It will most likely be a century of complexity. It must be a century of biological networks as well. If there is any area in which network thinking could trigger a revolution, I believe that biology is it.

THE FOURTEENTH LINK

Network Economy

TEN YEARS AGO AN EARLY and largely unknown Internet startup was desperately short of cash. As a manager for Time Warner, a member of the startup's directorial board saw these problems as an opportunity for the entertainment giant for which he worked. He therefore suggested to a Time Warner senior executive that they bail out the startup. For a mere $5 million the media conglomerate could have owned 11 percent of the company. This would have been petty cash for Time Warner and would have offered access to the Internet, at that time a brand new distribution channel. "If we did that," the senior executive replied, meaning that if he accepted the Internet as a viable distribution channel for Time Warner, "then everything we have done since 1923 would be thrown out the window."

He certainly was a terrible stock picker: Ten years later the $5 million investment would have been worth over $15 billion. The purchase would have altered history too. Indeed, a decade later Steve Case, the CEO of America Online (AOL), the once unknown Internet startup, and Jerry Levin, the chairman of Time Warner, announced the merger of the two companies at a Manhattan press conference. A few years earlier Time Warner could have easily digested the Internet startup. In 2000, however, it was AOL, a company that few had heard of a decade earlier, that swallowed the media giant.

Time Warner had content, and AOL had the means of delivering it to the consumer. Just before the collapse of the NASDAQ bubble in spring 2000, Jerry Levin was under pressure to go dot.com to regain Wall Street's attention, and Steve Case needed access to Time Warner's cable to get into your living room. Despite the very different cultures of the two companies, business analysts were eager to convince us that it was a match made in heaven. The same analysts had told us that the 1998 Daimler-Benz takeover of Chrysler also was a sound step for both companies. So was the fusion of the oil industry titans Exxon and Mobil in 1998, four months after another major acquisition in which Amoco was bought by British Petroleum. The list of attention-grabbing mergers and acquisitions does not end here, however. In 1998 alone Bell Atlantic paired up with GTE, SBC Communications bought Ameritech, BankAmerica joined up with NationsBank, Citicorp merged with Travelers Group.

Do these mergers make sense? Not if you listen to antiglobalization activists, who accuse big corporations of dictating everything from policy to fashion. They are unavoidable, however, if we view the economy as a complex network, whose nodes are companies and whose links represent the various economic and financial ties connecting them. Indeed, in a network economy the hubs must get bigger as the network grows. To satisfy their hunger for links, nodes of the business web learn to swallow the smaller nodes, a novel method unseen in other networks. As globalization pressures the nodes to grow bigger, mergers and acquisitions are a natural consequence of an expanding economy.

Motivated by the renaissance of networks in physics and mathematics, recently a number of new findings has documented the power of networks in everything from company structure to the marketplace. We have learned that a sparse network of a few powerful directors controls all major appointments in Fortune 1000 companies; a network of alliances determines the success in the biotech industry; the structure of the network within the firm is responsible for the organization's ability to adapt to rapidly changing market conditions; and strategies taking advantage of the network nature of the consumer base lead to phenomenal successes in marketing. As links and connections take over, understanding network effects become the key to survival in a rapidly evolving new economy.

1.

Regardless of industry and scope, the network behind all twentieth century corporations has the same structure: It is a *tree*, where the CEO occupies the root and the bifurcating branches represent the increasingly specialized and nonoverlapping tasks of lower-level managers and workers. Responsibility decays as you move down the branches, ending with the drone executors of orders conceived at the roots.

Despite its pervasiveness, there are many problems with the corporate tree. First, information must be carefully filtered as it rises in the hierarchy. If filtering is less than ideal, the overload at the top level, where all branches meet, could be huge. As a company expands and the tree grows, information at the top level inevitably explodes. Second, integration leads to unexpected organizational rigidity. A typical example comes from Ford's car factories, one of the first manufacturing plants to fully implement the hierarchical organization. The problem was that they got too good at it. Ford's assembly lines became so tightly integrated and optimized that even small modifications in automobile design required shutting down factories for weeks or months. Optimization leads to what some call *Byzantine monoliths*, organizations so overorganized that they are completely inflexible, unable to respond to changes in the business environment.

The tree model is best suited for mass production, which was the way of economic success until recently. These days, however, the value is in ideas and information. We have gotten to the point that we can produce anything that we can dream of. The expensive question now is, what should that be?

As companies face an information explosion and an unprecedented need for flexibility in a rapidly changing marketplace, the corporate model is in the midst of a complete makeover. This does not mean a superficial shift in the job description of a few individuals. It is a fundamental rethinking of how to respond to the new business environment in the postindustrial era, dubbed the information economy.

The most visible element of this remaking is a shift from a tree to a web or a network organization, flat and with lots of cross-links between the nodes. As valuable resources shift from physical assets to bits and information, operations move from vertical to virtual integration, the reach of businesses increasingly expands from domestic to global, the lifetime of inventories decreases from months to hours, business strategy changes from top-down to bottom-up, and workers transform into employees or free agents.

New products require new alliances both within and outside the company, demanding a new topology. To achieve this, layers of middle managers have been scrapped. Employees who previously played secondary roles are in charge of major products from one day to the next. Project teams, alliances within and outside the organization, and outsourcing proliferate. Therefore, companies aiming to compete in a fast-moving marketplace are shifting from a static and optimized tree into a dynamic and evolving web, offering a more malleable, flexible command structure. Those that resist this change could easily be forced to the periphery.

The internal remaking of the web within the firm is only one consequence of a network economy. Another is the realization that companies never work alone. They collaborate with other institutions, adapting business practices proved successful in other organizations. The crucial high-level connection to the rest of the corporate world is often maintained by the CEO and the board of directors. As we will see next, network effects play a fundamental role in these interactions.

2.

"I want to say to you absolutely and unequivocally that Ms. Lewinsky told me in no uncertain terms that she did not have a sexual relationship with the President," read Vernon Jordan at a hastily convened press conference in the midst of the Clinton-Lewinsky scandal. But he soon was to "pull off some of the fanciest footwork of his career—dancing out of the box that he put himself in," according to *Time* magazine's Eric Pooley, as everyone pressed him for a satisfactory explanation for the four meetings and seven phone calls Jordan had with the former White House intern, trying to arrange a job for her at one of several major companies.

Jordan's role in finding Monica Lewinsky a corporate job was no surprise to Washington insiders. His inability to steer the attention away from himself was something new, however. An effective civil rights leader in the 1970s, Jordan was shot in the back in 1980 by a white supremacist, who settled on him after learning that Jesse Jackson, whom he really wanted to kill, was out of town. Jordan carefully had avoided the spotlight ever since, becoming the most powerful unknown in D.C., a rarely heard or seen top deal maker and superlawyer in Washington's media-fixated crowd. As Pooley wrote in *Time*, Jordan "earns $1 million a year from a law practice that requires him to file no brief and visit no courtroom, because his billable hours tend to be logged in posh restaurants, on cellular telephones, in the tufted-leather backseats of limousines—making a deft introduction here, nudging a legislative position there, ironing out an indelicate situation before it makes the papers."

Uncharacteristically, Jordan found *himself* in the papers all over the nation in 1998, his meetings and phone calls being scrutinized by everyone from the media to independent counsel Kenneth Starr. He emerged as a prominent node in the entangled web of the Clinton-Lewinsky scandal, often dubbed the Six Degrees of Monica.

Jordan was not a newcomer to small worlds. He acquired his unique status as a consummate Washington insider by successfully surfing one of the most influential small-world networks in the American economy, the corporate web. During the years preceding the Clinton-Lewinsky scandal and the Clinton presidency, Jordan became the most central director of the small corporate elite running the Fortune 1000 corporate world.

The board of directors, a group of about a dozen individuals, holds unusual power in overseeing a company's future. It is responsible for all major decisions, from ousting poorly performing CEOs to approving major mergers and acquisitions. Therefore, corporations make all efforts to recruit well-connected and experienced directors. Successful CEOs, lawyers, and politicians are frequently sought after, being courted for directorship on several boards at the same time.

Despite concerns that directors serving on a large number of boards cannot possibly find the time to do justice to all of them, most companies want their directors to have experience on other boards. As directors apply the knowledge and experience they acquired on one board to bear on questions faced by another, this interlocked network of board members plays a crucial role in spreading corporate practices and maintaining the political and economic clout of big corporations.

Thanks to the important role boards play in shaping the landscape of American corporate life, the web of directors has often been scrutinized in business literature. But only recently, with the advent of methods to analyze complex networks, have we started to understand to what degree the power of this web is rooted in its interlocked topology.

In the director network each node is a board member linked to directors serving on the same board. With thousands of companies, each with about a dozen or so directors, this is a rather large web. Gerald F. Davis, Mina Yoo, and Wayne E. Baker, from the University of Michigan Business School, recently studied the most influential component of this web, focusing on the network of Fortune 1000 companies, made up of 10,100 directorships held by 7,682 directors. If each director were to serve on one board only, the network would be broken into tiny, fully connected circles, each the size of a single board. This is not the case, however. While 79 percent of directors serve on only one board, 14 percent serve on two, and about 7 percent serve on three or more. The measurements indicated that these few overlapping directors create a small-world network with five degrees of separation. Indeed, the distance between any two directors belonging to the major cluster, which contains 6,724 directors, was 4.6 handshakes on average.

The small-world nature of the director web is due to the 21 percent of directors who serve on more than one board, since they are the ones who hold this complex network together. Of these, Vernon Jordan plays a very special role. With membership on ten boards, in which he regularly meets 106 other Fortune 1000 directors, Jordan is the most central director of the corporate elite, within three handshakes from most other directors.

3.

Jordan's career offers a vivid demonstration of how the interlocked, small-world nature of corporate directorships determines most major appointments in corporate life. Indeed, in most cases when Jordan joined a board, he already knew at least one director from his service on other boards. In the early seventies, as president of the National Urban League, the influential civil rights organization, Jordan repeatedly called for the inclusion of blacks in the powerful corporate elite. In 1972 John Brooks, the chairman of Celanese Corporation, a diversified manufacturer of chemicals, told him, "I think you ought to put your money where your mouth is. . . . You're talking about blacks on the board of directors. Why don't you come on the board at Celanese?"

Soon after joining the board of Celanese, Jordan received two calls inviting him to join the boards of both Marine Midland Bank and Bankers Trust. Undecided as to which he should accept, Jordan called John Brooks for advice. "You don't have a choice. It's Bankers Trust," came the short reply. When Jordan asked why, Brooks answered simply, "How do you think you got nominated to be on the Bankers Trust board? I am on the board. I nominated you." At Bankers Trust Jordan served together with William M. Ellinghaus, who held a directorship at JC Penney as well. A year later Jordan was invited to serve on the board of JC Penney.

Three years later Jordan asked Peter McCullough, the CEO of Xerox, to be the corporate chairman of the National Urban League. He accepted with a condition: "I'll be your corporate chairman if you come on the Xerox board." Jordan agreed. Three years after becoming a Xerox director, Jordan was invited to the board of American Express, where two other Xerox directors already served. It comes as no surprise that in 1980 Jordan joined the board of RJ Reynolds. Indeed, the CEO of Celanese and another JC Penney board member both served on the RJ Reynolds board, and Jordan had close links to the RJ Reynolds CEO as well, who was a fellow director on the Celanese board.

Prior acquaintanceship allows directors to vouch for prospective recruits. Therefore, the small-world dynamics help the creation of a powerful "old boy network," or corporate elite, that has unparalleled influence in economic and political life. Jordan's current job at Akin, Gump, Strauss, Hauer & Feld, one of the biggest law practices in Washington, can be also traced back to this old boy network: Robert S. Strauss, the partner responsible for recruiting Jordan, was a fellow director on the Xerox board.

Jordan's path is by no means unique. Network effects are known to be present in all industries. For example, in Silicon Valley the extensive movements of labor between companies create dense personal intercompany links. These subtle social networks are extensively utilized for hiring new employees and attracting managers. Since current employees can vouch for their social links, just as directors do for fellow board members, employees hired through social networks quit less frequently and perform better than those recruited otherwise.

The intricate and interlocked nature of board directorships and Silicon Valley employees provides just two examples of the complex social and power networks behind the U.S. economy. But to comprehend how an economy truly works, we need to understand how corporations and other economic institutions run by these highly connected directors interact with each other.

4.

Although universities and their spin-offs, small biotech companies, have been recently the driving force behind the development of new drugs, the cash and experience needed to launch large-scale clinical trials and the worldwide marketing channels continue to be located in large chemical and pharmaceutical companies. Because the development and marketing of a new drug can cost anywhere from $150 million to $500 million, the different players of this field, ranging from universities and research labs to government agencies, chemical and pharmaceutical companies and venture capital firms, have been forced to form strategic partnerships. These alliances, together with the relatively young age of the biotech industry, offer an unusually well documented case of network formation, allowing us to follow and understand the emergence of networks in economic systems.

From its early days the biotech industry displayed the essential attributes of a growing network. This growth was captured in a dynamic graph developed by Walter W. Powell, Douglas White, and Kenneth W. Koput, depicting the biotech network at different stages of its evolution between 1988 and 1999. In 1988, representing the early days of the industry, there were far fewer links than nodes: Seventy-nine organizations connected by only thirty-one links. According to the famous Erdős-Rényi prediction, the network should have been broken into many tiny clusters. In reality, however, the nodes formed two major components, one with twenty-seven and the other with four organizations. That is, none of the thirty-one links was wasted—each of them contributed to a major component developing around a few biotech companies, leading to a level of connectedness that could not emerge in a random network. A few hubs visible already at this early stage were the first-mover biotech companies, such as Centocor, Genzyme, Chiron, Alza, and Genentech. Without them the biotech network would have broken into many tiny disconnected nodes.

But the existence of a few companies with a large number of partnerships, resembling hubs, is not enough for us to identify the nature of the network. For this we have to analyze the degree distribution, a study recently performed by two economists, Massimo Riccaboni and Fabio Pammolli, both from the University of Siena, working with physicist Guido Cardarelli from La Sapienza University in Rome, Italy. Their study was based on data collected by the Pharmaceutical Industry Database, hosted by the University of Siena, which provides information for 3,973 research and development agreements between 1,709 firms and institutions. The analysis indicates that the hubs noticed by Powell, White, and Koput are not accidental but are rooted in the scale-free nature of the network behind the pharmaceutical industry. Indeed, the number of companies that entered in partnership with exactly k other institutions, representing the number of links they have within the network, followed a power law, the signature of a scale-free topology. A hierarchy of well-connected large corporations brought together a large number of small companies, seamlessly integrating all players into an evolving scale-free economy.

As research, innovation, product development, and marketing become more and more specialized and divorced from each other, we are converging to a network economy in which strategic alliances and partnerships are the means for survival in all industries. The interfirm linkages of suppliers and subcontractors are well documented in southwestern Germany and north central Italy; Japanese business has long relied on interfirm collaborations to diffuse responsibilities for technological innovations; the Korean business model marries a whole array of diverse companies under the umbrella of large conglomerates; Silicon Valley regularly takes advantage of technology transfers by pairing up startups with established companies. These fluid alliances, which are periodically renegotiated as the marketplace shifts or the focus of the participants changes, offer a glimpse of the future of the world's business environment.

5.

Despite the important role these interfirm alliances play in the economy, economic theory pays surprisingly little attention to networks. Until recently economists viewed the economy as a set of autonomous and anonymous individuals interacting through the price system only, a model often called the *standard formal model* of economics. The individual actions of companies and consumers were assumed to have little consequence on the state of the market. Instead, the state of the economy was best captured by such aggregate quantities as employment, output, or inflation, ignoring the interrelated microbehavior responsible for these aggregate measures. Companies and corporations were seen as interacting not with each other but rather with "the market," a mythical entity that mediates all economic interactions.

In reality, the market is nothing but a directed network. Companies, firms, corporations, financial institutions, governments, and all potential economic players are the nodes. Links quantify various interactions between these institutions, involving purchases and sales, joint research and marketing projects, and so forth. The *weight* of the links captures the value of the transaction, and the direction points from the provider to the receiver. The structure and evolution of this weighted and directed network determine the outcome of all macroeconomic processes.

As Walter W. Powell writes in *Neither Market nor Hierarchy: Network Forms of Organization*, "in markets the standard strategy is to drive the hardest possible bargain on the immediate exchange. In networks, the preferred option is often creating indebtedness and reliance over the long haul." Therefore, in a network economy, buyers and suppliers are not competitors but partners. The relationship between them is often very long lasting and stable.

The stability of these links allows companies to concentrate on their core business. If these partnerships break down, the effects can be severe. Most of the time failures handicap only the partners of the broken link. Occasionally, however, they send ripples through the whole economy. As we will see next, macroeconomic failures can throw entire nations into deep financial disarray, while failures in corporate partnerships can severely damage the jewels of the new economy.

6.

On February 5, 1997, Somprasong Land, a Thai property development company, failed to pay interest of $3.1 million on Euro-convertible debt. In a globalized economy where trillions of dollars change hands daily, this is petty cash. Not surprisingly, the event easily evaded the attention of the average investor. Unnoticed by most, this single failure was nevertheless the spark that led to the melting of the world's financial architecture.

A month later the Thai government made the first in a series of desperate attempts to save the country's economy from imminent collapse, announcing that it would buy $3.9 billion in bad property debt from financial institutions. A few days later it reneged on its promise, a move that some financial experts took as a sign of stability. The

International Monetary Fund's managing director, Michel Camdessus, who was later criticized for his organization's role in the Asian financial meltdown, said, "I don't see any reason for this crisis to develop further."

Subsequent events proved him wrong. Two weeks later the financial sector was trembling in Malaysia, prompting its central bank to restrict loans. At the same time, Sammi Steel, the main firm of Korea's twenty-sixth largest conglomerate, sought court receivership, the first step toward bankruptcy. In May, Japan hinted that it would raise interest rates to stop the decline of the yen (which never happened), triggering a global sell-off of Southwest Asian currencies and shaking the local stock markets. A week later Thailand failed to save its largest finance company, Finance One, which effectively went bankrupt. The event triggered a strong speculative attack on Thailand's currency, the baht, which, despite repeated promises to the contrary by the government, was abandoned on July 2.

The cascading failures of companies and financial institutions in Thailand, Indonesia, Malaysia, Korea, and the Philippines would take hundreds of pages to fully document. So would the chronicle of finger-pointing, including such highlights as Malaysian Prime Minister Mahathir Mohamad's bitter attack on "rogue speculators," which culminated in a talk given to the IMF/World Bank annual conference in which he called currency trading immoral. George Soros, the prominent international financier, responded a day later, "Dr. Mahathir is a menace to his own country."

Some economists blamed the "structural and policy distortions in the countries of the region" for the financial meltdown. Yet President Clinton and his economic team in the economic report of the president to the Congress in 1999 maintained that the crisis "was not due to problems with the economic fundamentals." Less than a year after the events, Paul Krugman, professor of economics and international affairs at Princeton, summarized the overall feeling: "It seems safe to say that nobody anticipated anything like the current crisis in Asia." A few small, localized financial difficulties had set off a chain reaction of failures that swept across national boundaries, creating a huge currency devaluation and stock market crashes from Asia to South America. It eventually caused the single biggest point loss ever of the Dow Jones industrial average, which tumbled 554.26 points on October 27, 1997.

How could the failure of a large but far from dominant property development company shake the world's largest stock market and keep the president of the "world's strongest nation" explaining even two years after? If we view the economy as a highly interconnected network of companies and financial institutions, we can begin to make sense of these events. In such networks the failure of a node has little effect on the system's integrity. Occasionally, however, the breakdown of some well-selected nodes sets off a cascade of failures that can shake the whole system.

The Asian crisis was a large-scale example of a cascading financial failure similar to those we discussed in Chapter 9, a natural consequence of connectedness and interdependency. It was not the first, however: South America and Mexico had experienced similar cascading failures two years earlier. It is surely not the last either, despite all the measures banks and governments seem to have taken to avoid it.

These events cannot be explained within a framework in which all organizations interact with a mythical market only. Cascading failures are a direct consequence of a network economy, of interdependencies induced by the fact that in a global economy no institution can work alone. Understanding macroeconomic interdependencies in terms of networks can help us to foresee and limit future crises. Thinking networks can teach us to monitor the path of the damage and to set firewalls by identifying and strengthening the nodes that can stop the spread of macroeconomic fires.

We should not let ourselves believe that such cascading failures as the Asian crisis and its Latin American counterparts are the side effects of the unstable financial systems of rapidly developing nations. Established economies, such as the United States', that have the cash and the expertise to root out such failures before they turn global aren't immune to cascading failures. Vulnerabilities related to interconnectivity exist in stable economies as well, as the burst of the dot.com bubble illustrates.

7.

In late 1999, Compaq's Pocket PC became the company's biggest hit. As discussed by a recent *Strategy & Business* study, demand for the device outpaced supply twenty-five times, making some Compaq executives dream that, with support and accessories, the handheld devices could soon offer a bigger market than traditional PCs. Then problems started surfacing.

Compaq, Cisco Systems, and several other companies are leaders of a new business strategy: outsourcing. Cisco, which not long ago was poised to become the first trillion-dollar company, is the driving force behind this trend. It reached a 30 to 40 percent annual revenue growth with a novel and aggressive approach to manufacturing: It didn't build anything that it sold. Rather, it established strong ties to a large number of manufacturers who built and assembled the pieces sold under Cisco's logo. Compaq and many others followed suit.

Outsourcing requires a tight integration of suppliers, making sure that all pieces arrive just in time. Therefore, when some suppliers were unable to deliver certain basic components like capacitors and flash memory, Compaq's network was paralyzed. The company was looking at 600,000 to 700,000 unfilled orders in handheld devices. The $499 Pocket PCs were selling for $700 to $800 at auctions on eBay and Amazon.com. Cisco experienced a different but equally damaging problem: When orders dried up, Cisco neglected to turn off its supply chain, resulting in a 300 percent ballooning of its raw materials inventory.

The final numbers are frightening: The aggregate market value loss between March 2000 and March 2001 of the twelve major companies that adopted outsourcing—Cisco, Dell, Compaq, Gateway, Apple, IBM, Lucent, Hewlett-Packard, Motorola, Ericsson, Nokia, and Nortel—exceeded $1.2 trillion. The painful experience of these companies and their investors is a vivid demonstration of the consequences of ignoring network effects. A *me* attitude, where the company's immediate financial balance is the only factor, limits network thinking. Not understanding how the actions of one node affect other nodes easily cripples whole segments of the network.

Experts agree that such rippling losses are not an inevitable downside of the network economy. Rather, these companies failed because they outsourced their manufacturing without fully understanding the changes required in their business models. Hierarchical thinking does not fit a network economy. In traditional organizations, rapid shifts can be made within the organization, with any resulting losses being offset by gains in other parts of the hierarchy. In a network economy each node must be profitable. Failing to understand this, the big players of the network game exposed themselves to the risks of connectedness without benefiting from its advantages. When problems arose, they failed to make the right, tough decisions, such as shutting down the supply line in Cisco's case, and got into even bigger trouble.

At both the macro- and the microeconomic level, the network economy is here to stay. Despite some high-profile losses, outsourcing will be increasingly common. Financial interdependencies, ignoring national and continental boundaries, will only be strengthened with globalization. A revolution in management is in the making. It will take a new, network-oriented view of the economy and an understanding of the consequences of interconnectedness to smooth the way.

8.

Sabeer Bhatia did not know how to sell a company. But having been born and raised in India, he did know how to buy onions. You have to negotiate. Now he had a very hot onion to sell. He and his partner, Jack Smith, on July 4, 1996, launched a service offering nothing but e-mail—free to anybody in the world. They named it Hotmail. By year's end they had signed up a million customers, each of whom view daily the banner ads displayed on their e-mail account, Hotmail's main source of revenue. When Microsoft came knocking a year later, nearly 10 million users had Hotmail accounts. Bhatia was only twenty-eight when, after touring all twenty-six buildings at Microsoft's Redmond, Washington, empire and shaking hands with Bill Gates, he was ushered into a room packed with twelve Microsoft negotiators. They offered him $160 million. "I'll get back to you," he said, and walked away.

Currently Hotmail has about a quarter of all e-mail accounts. It is the biggest e-mail service provider in Sweden and India, countries in which it has never advertised. Microsoft eventually paid $400 million for the company, which a year later, before the burst of the dot.com bubble, was worth $6 billion.

How did an underfunded startup sign up a quarter of all e-mail users? The answer is simple: They exploited the power of networks, using a hot new marketing technique called viral marketing. Viral marketing works on the same principle that allowed Love Bug to circle the globe in a few hours. The computer virus reached everybody by looking up the e-mail list you store in your Microsoft Outlook program, sending a copy of itself to each address. Thanks to a similar innovation, Hotmail users voluntarily offer the same service.

Tim Draper, from the Draper, Fisher and Jurvetson venture capital firm, after providing $300,000 seed money to launch Hotmail, persuaded Bhatia and Smith to add an extra line at the end of each email: "Get Your Private, Free Email at http://www.hotmail.com." Therefore, whenever Hotmail users send e-mails to their friends, they advertise and endorse the company. The news about Hotmail travels on a scale-free network, utilizing exactly the same routes that helped the spread of Love Bug. Because the critical threshold for innovation spreading vanishes on such networks, it was likely that Hotmail would succeed. It was unexpected and surprising, however, how fast and to what degree it did.

What is the source of Hotmail's phenomenal success? The answer is partially contained in the Trieste study discussed in Chapter 10. Innovations and products with a higher spreading rate have a higher chance of reaching a large fraction of the network. Hotmail enhanced its spreading rate by eliminating the adoption threshold individuals experience. First, it is free; thus you do not have to think about whether you are making a wise investment. Second, the Hotmail interface makes it very easy to sign up. In two minutes you have an account; thus there is no time investment. Third, once you sign up, every time you send an e-mail, you offer free advertisement for Hotmail. Combine these three features, and you get a service that has a very high infection rate, a built-in mechanism to spread. Traditional marketing theories will tell you that the combination of free service, low learning path, and rapid reach through consumer marketing has put the product above the threshold, and that is why it reached everybody. Based on our new understanding of diffusion in complex networks, we now know that this is only partially correct. It is true that you have a very high rate of spread. But you have no threshold either. Products and ideas spread by being adapted by hubs, the highly connected nodes of the consumer network.

Can Hotmail be replicated? Don't bet on it. Take for example EpidemicMarketing.com, a company that spent $2.1 million on a thirty-second Super Bowl advertisement in 2000, dreaming big to exploit the power of networks. In the Super Bowl ad a man visits a public restroom and receives a tip from the washroom attendant, instead of tipping the attendant as is customary. As was so cleverly expressed in their commercial, Epidemic planned to reward people for doing things they do every day. Their business model was to pay consumers to attach links to Internet businesses on their outgoing e-mail. Therefore, information about a company or promotion was expected to spread largely through word of mouth, replicating the phenomenal success of Hotmail. The model was missing a crucial element of viral marketing, however: Your friend had little interest in passing on the link to his or her acquaintances. It comes as no surprise, therefore, that Epidemic closed its doors and laid off its sixty-person staff in June 2000 after burning through the $7.6 million it raised.

Hotmail demonstrates the power of consumer networks. Some products do not need expensive telemarketing or TV and newspaper ads to prevail. They simply spread by word of mouth like a virus. Though it may not work for all products, throwing in elements of viral marketing could enhance just about all sales. Yet Epidemic's failure indicates that Hotmail cannot be easily copied. Instead, Hotmail's experience should be the starting point for new marketing approaches, combining traditional strategies with a better understanding of network effects.

9.

Network effects proliferate in the business world. We saw Vernon Jordan successfully surf the complex corporate network, becoming an influential member of the corporate elite. We saw Hotmail take advantage of the scale-free nature of the consumer network to become the biggest e-mail provider worldwide. The list does not stop here. Motivated by the evolving marketplace, an array of new companies have lately vowed to put network thinking at the core of their business models. Their record is mixed at best.

Take for example SixDegrees.com, a New York–based startup that asked its members to submit the names of their friends, inviting them to join too. If they enrolled, they also submitted the names of *their* friends. Step by step SixDegrees acquired a detailed map of the social network around each of its members, allowing them to reach everybody two links away from them. This consumer-driven viral marketing allowed SixDegrees to sign up over 3 million consumers. Yet the startup closed its doors on December 30, 2000, failing to turn six degrees into a viable business plan.

The burst of the dot.com bubble is often attributed to the one-dimensional thinking of many Internet enthusiasts. Most startups were based on the simple philosophy that offering things online was sufficient to replicate the success stories of the new economy. Yet, apart from a few early starts, such as Amazon.com, AOL, or eBay, most failed. The real legacy of the Internet is not the birth of thousands of new online companies but the transformation of existing businesses. We can see its signature on everything from mom-and-pop stores to large multinational agglomerates.

Networks do not offer a miracle drug, a strategy that makes you invincible in any business environment. The truly important role networks play is in helping existing organizations adapt to rapidly changing market conditions. The very concept of network implies a multidimensional approach.

The diversity of networks in business and the economy is mind-boggling. There are policy networks, ownership networks, collaboration networks, organizational networks, network marketing—you name it. It would be impossible to integrate these diverse interactions into a single all-encompassing web. Yet no matter what organizational level we look at, the same robust and universal laws that govern nature's webs seem to greet us. The challenge is for economic and network research alike to put these laws into practice.

THE LAST LINK

Web Without a Spider

BY MARCH 1998, when in an unusual move I invited Réka Albert to lunch, she was only a year and a half into her graduate studies but had enough publications to receive a Ph.D. One of her papers, on granular media and sand castles, was featured on the cover of *Nature* and *Science News*, and the preliminary results of her current projects were promising, as well. So the purpose of this lunch defied all wisdom: I wanted to persuade her to give up the research she had been so good at and start something entirely different. I told her about my dream to study networks.

Four years earlier, in the fall of 1994, with a fresh doctorate in theoretical physics, I had started as a postdoc in the legendary corporate ivory tower of IBM, the T. J. Watson Research Center in Yorktown Heights, New York. Four months into my job there, perhaps touched by the spirit of the place, I checked out from the library a general-audience book on computer science to read over the Christmas break. As I immersed myself in algorithms, graphs, and Boolean logic, I started to sense how little was known about networks in general. All my readings told me that the millions of electric, telephone, and Internet cables cramped under the pavement in Manhattan formed a fundamentally random network. The more I thought about it, the more I was convinced that there must be some organizing principles governing the complex webs around us. Dreaming of identifying some signature of order, I started to study network theory, beginning with the classical works of Erdős and Rényi. Before I left IBM in the fall of 1995 for a faculty position in physics at University of Notre Dame, I had submitted my first research paper about complex networks.

At Notre Dame, I tried with little success to contact search engines for data on the Web's topology. Under pressure to publish and obtain grants, I gradually replaced networks with safer and more conventional research. By the beginning of 1998, however, I was ready to return to thinking about nodes and links. Now I was asking one of my best students to drop everything she was doing and join me on that risky journey. I could offer her little encouragement at that time. I had to tell her that my only paper about networks had been rejected by four journals and never been published. I told her she was risking a sudden end to the success story she was part of so far. But I also told her that sometimes we should be ready to take risks. In my view, networks were worth the try.

In 1994, or even in early 1998, nobody could have anticipated the flood of discoveries that would completely reshape our understanding of our interconnected world in the following years. At that lunch with Albert when I made my pitch for networks, I could not tell her about small worlds. Not even in my wildest dreams could I conjure power laws or scale-free networks. I could not talk about error-and-attack tolerance either, since these were nonissues in network research at that time. In fact, every question worth studying that I could tell her about has since been proven ill-founded or simply irrelevant.

It was Hawoong Jeong's robot that forced us to think outside the box. Jeong joined my research group as a postdoctoral researcher in August 1998, five months after Albert and I took up networks as a research topic. Recently graduated from Korea's prestigious Seoul National University, his fascination with and knowledge of computers were prodigious. One day, after a late night discussion, I casually asked him if he would be able to build a robot to map out the World Wide Web. He made no promises. But a month later his robot was busy carrying home the nodes and links. By that time we were somewhat familiar with the classical literature dealing with random graphs and networks. Thus it was immediately clear that the power laws seen by the robot represented a serious deviation from everything then known about networks. It was only after the construction of the scale-free model, however, that we fully understood how different real webs are from the random universe Erdős and Rényi depicted.

Today we know that, though real networks are not as random as Erdős and Rényi envisioned, chance and randomness do play an important role in their construction. Real networks are not static, as all graph theoretical models were until recently. Instead, growth plays a key role in shaping their topology. They are not as centralized as a star network is. Rather, there is a hierarchy of hubs that keep these networks together, a heavily connected node closely followed by several less connected ones, trailed by dozens of even smaller nodes. No central node sits in the middle of the spider web, controlling and monitoring every link and node. There is no single node whose removal could break the web. A scale-free network is a web without a spider.

In the absence of a spider, there is no meticulous design behind these networks either. Real networks are self-organized. They offer a vivid example of how the independent actions of millions of nodes and links lead to spectacular emergent behavior. Their spiderless scale-free topology is an unavoidable consequence of their evolution. Each time nature is ready to spin a new web, unable to escape its own laws, it creates a network whose fundamental structural features are those of dozens of other webs spun before. The robustness of the laws governing the emergence of complex networks is the explanation for the ubiquity of the scale-free topology, describing such diverse systems as the network behind language, the links between the proteins in the cell, sexual relationships between people, the wiring diagram of a computer chip, the metabolism of the cell, the Internet, Hollywood, the World Wide Web, the web of scientists linked by coauthorships, and the intricate collaborative web behind the economy, to name only a few.

1.

One of the most fascinating aspects of the birth of a new science is the new language it creates, allowing us to casually converse about ideas and issues that we were struggling to describe before. The renaissance of network theory has done this for our interconnected world. The connectors of society, the stars of Hollywood, and the keystone species of an ecosystem are suddenly only manifestations of a single reality, their perceived importance within their environment attributable to their status as hubs within their respective networks. Network thinking is poised to invade all domains of human activity and most fields of human inquiry. It is more than another helpful perspective or tool. Networks are by their very nature the fabric of most complex systems, and nodes and links deeply infuse all strategies aimed at approaching our interlocked universe.

A dramatic example of the pervasiveness of this new language came after September 11, 2001, when networks acquired a meaning previously unfamiliar to most of us. Most of what led to the tragedy make perfect sense from a network perspective. Al Qaeda, the terrorist network held responsible for the attacks, was not created in seven days. Driven by religious beliefs and impatience with the existing social and political order, thousands were drawn to the radical organization over several years. The network expanded one node at a time, taking on all the characteristics of a web without a spider. Indeed, al Qaeda failed to turn into the hub-and-spoke network that offers a central leader control over all details. It avoided the tree structure as well, the chain of command characterizing the military and twentieth-century corporations. Rather, it evolved into a self-organized spiderless web in which a hierarchy of hubs kept the organization together.

After September 11, Valdis Krebs, a management consultant who normally uses network theory to analyze corporate communications, assembled a map of the nineteen hijackers aboard the four planes involved in the attacks and the fifteen people whom authorities claimed to have been connected to them. Krebs carefully entered all publicly disclosed contacts between these thirty-four individuals, weighting the links based on the known closeness of the relationship. The obtained web is extremely revealing for anybody who wants to understand the inner workings of the deadly cell that carried out the attacks. It offers few surprises to those familiar with the shape of real networks. Mohamed Atta, the purported mastermind of the attacks, is the most connected node indeed. Yet, he had direct contact with only sixteen of the twenty-three nodes. He is closely trailed by Marwan Al-Shehhi, the second most connected node, with links to fourteen nodes. As we go down the list, we encounter numerous nodes poor in links, the peripheral soldiers of the deadly organization.

The map also shows that, despite his central role, taking out Atta would not have crippled the cell. The rest of the hubs would have kept the web together, possibly carrying out the attack without his help. Many suspect that the structure of the cell involved in the September 11 attack characterizes the whole terrorist organization. Because of its distributed self-organized topology, Al Qaeda is so scattered and self-sustaining that even the elimination of Osama bin Laden and his closest deputies might not eradicate the threat they created. It is a web without a true spider.

Today the world's most dangerous aggressors, ranging from al Qaeda to the Colombian drug cartels, are not military organizations with divisions but self-organized networks of terror. In the absence of familiar signs of organization and order, we often call them "irregular armies." Yet by doing so we again equate complexity with randomness. In reality, terrorist networks obey rigid laws that determine their topology, structure, and therefore their ability to function. They exploit all the natural advantages of self-organized networks, including flexibility and tolerance to internal failures. Unfamiliarity with this new order and a lack of language for formalizing our experience are perhaps our most deadly enemies.

To be sure, the battle against al Qaeda can and will be won by crippling the network, either by removing enough of its hubs to reach the critical point for fragmentation or by draining its resources, preparing the groundwork for cascading internal failures. Yet, collapsing al Qaeda will not end the war. Other networks with similar scope and ideology will no doubt take its place. Bin Laden and his lieutenants did not invent terrorist networks. They only rode the rage of Islamic militants, exploiting the laws of self-organization along their journey. If we ever want to win the war, our only hope is to tackle the underlying social, economic, and political roots that fuel the network's growth. We must help eliminate the need and desire of the nodes to form links to terrorist organizations by offering them a chance to belong to more constructive and meaningful webs. No matter how good we become at winning each net battle, if we are unable to inhibit the desire for links, the prerequisite for the formation of these deadly self-organized webs, the net war will never end.

2.

On June 23, 1995, the *New York Times* carried a large photograph of the German parliament, the century-old Reichstag, on its cover. This was five years after German reunification and almost exactly four years after the Bundestag, sitting in Bonn, voted to make Berlin the capital of the united Germany once again. Yet, politics and the collapse of communism had little to do with the renewed worldwide attention to the Reichstag. The real attraction for the 5 million visitors who flooded to Berlin during the coming two weeks was the fact that none of them could actually spot even a square inch of the building. The Reichstag's signature sober gray walls, the dark and quiet witnesses of a century of tumultuous German history, were all invisible. This ultimate symbol of power was wrapped in an aluminum-colored fabric, from its stairs to its flag post, transforming it into a monumental piece of public art. Over a million square feet of thickly woven polypropylene fabric held together by 5,000 feet of blue rope covered every square inch of the structure, offering one of the most magnificent artistic spectacles of our time.

The portfolio of the Bulgarian-born artist Christo and his partner, the French artist Jeanne-Claude, includes such monumental works as the Wrapped Pont Neuf, which covered the famous Parisian bridge with a yellow drapery, and the magnificent Surrounded Islands, for which they placed six million square feet of pink fabric around eleven islands in Biscayne Bay, Miami, Florida. In many ways the Wrapped Reichstag was the culmination of their decades-long wrapping activity. Yet, it would be simplistic to perceive the artists simply as wrappers of buildings, bridges, and other objects. Their work has a powerful philosophy: "revelation through concealment." By hiding the details they allow us to focus entirely on the form. The wrapping sharpens our vision, making us more aware and observant, turning ordinary objects into monumental sculptures and architectural pieces.

In a sense we approached the world in this book following the spirit of Christo and Jeanne-Claude. To look at the networks behind such complex systems as the cell or the society, we concealed all the details. By seeing only nodes and links, we were privileged to observe the architecture of complexity. By distancing ourselves from the particulars, we glimpsed the universal organizing principles behind these complex systems. Concealment revealed the fundamental laws that govern the evolution of the weblike world around us and helped us understand how this tangled architecture affects everything from democracy to curing cancer.

Where do we go from here? The answer is simple. We must remove the wrapping. The goal before us is to understand complexity. To achieve that, we must move beyond structure and topology and start focusing on the dynamics that take place along the links. Networks are only the skeleton of complexity, the highways for the various processes that make our world hum. To describe society we must dress the links of the social network with actual dynamical interactions between people. To understand life we must start looking at the reaction dynamics along the links of the metabolic network. To understand the Internet, we must add traffic to its entangled links. To understand the disappearance of some species in an ecosystem, we have to acknowledge that some prey are easier to catch than others.

In the twentieth century we went as far as we could to uncover and describe the components of complex systems. Our quest to understand nature has hit a glass ceiling because we do not yet know how to fit the pieces together. The complex issues with which we are faced, in fields from communication systems to cell biology, demand a brand new framework. Embarking on the journey ahead without a map would be hopeless. Fortunately the ongoing network revolution has already provided many of the key maps. Though there are still many "dragons" ahead, the shape of a new world has become discernible, continent by continent. Most important, we have learned the laws of web cartography, allowing us to draw new maps whenever we are faced with new systems. Now we must follow these maps to complete the journey, fitting the pieces to one another, node by node and link by link, and capturing their dynamic interplay. We have ninety-eight years to succeed at this, and make the twenty-first the century of complexity.

AFTERLINK

Hierarchies and Communities

WHILE SIX DEGREES AND SMALL WORLDS have been with us since the 1960s, it wasn't until the late 1990s that we started to uncover the intricate organizing principles shaping such complex systems as the cell or the Internet. The discovery of scale-free networks induced a paradigm shift: They taught us that the many complex webs surrounding us are far from random, but are characterized by the same robust and universal architecture. I am repeatedly asked a few basic questions when I lecture about networks: Why did it take this long? Why did we have to wait until 1999 to discover the impact of hubs and power laws on the behavior of complex networks? The answer is simple: We lacked a map. The few network maps available for study before the late 1990s had a few hundred nodes at most. The enormous World Wide Web offered the first chance to examine the intricate anatomy of large complex systems and established the presence of power laws. As other large maps followed, we gradually understood that most networks of practical interest, from the language to the sex web, are shaped by the same universal laws and therefore share the same hub-dominated architecture.

This paradigm shift was so profound and took place so fast that we still struggle to fully comprehend its implications, forcing us to systematically revisit the role of networks in most fields. Hubs have prompted epidemiologists to seek new strategies to halt the AIDS pandemic or contain a potential small pox outbreak. In an era when national discourse is dominated by terrorism, the Internet's robustness and fragility drive research toward more secure communication systems. Network thinking has opened up new research fields, from systems biology to genetic networks, the driving forces of the rapidly unfolding postgenomic revolution. In the past two years alone scale-free networks have been utilized in drug discovery and cancer research; to characterize the large-scale structure of language and redesign Internet network generators; to design efficient search algorithms for distributed databases to facilitate finding detailed information on the World Wide Web quickly; to investigate Gnutella, the distributed Napster alternative; and to quantify within the Marvel Universe the role of such comics characters as Spider-Man or Captain America.

The jewels of network research, however, have been hidden beneath a mathematical language impenetrable to all but a few researchers. Computer scientists and physicists, conversant with the mathematical language of networks, are rapidly taking advantage of the unwinding network revolution. Fields accustomed to a less mathematical discourse remain largely unaffected.

Two thousand and two appears to have been a turning point, the opening of a wide discourse about networks. Among the hundreds of e-mails in response to *Linked*, I have heard from military strategists thinking about the role of hubs in security issues and terrorism, from a venture capitalist who launched a network-based start-up company after reading the book, from an activist who believes that charting our links could help us establish colonies on the moon, from a CEO of an Internet networking company who shared his experience with hubs and nodes, from a scholar of the ancient Islamic world who explored the web of powerful mullahs at the turn of the previous millennium. *Linked* has been particularly embraced by business and biology. The strong response from such different communities is testimony to a powerful trend rapidly turning the twenty-first century into the century of connectedness. As the immunologist Heinz-Günter Thiele recently remarked, we "cannot any longer imagine modern life and science without notions of networks and network theories." By chronicling the process through which we acquired the language of networks, *Linked* has helped turn network thinking into a rapidly diffusing virus, spreading on the very networks it describes. Yet, despite the many small worlds around us, ideas rarely take the shortest route. One thing is clear, however: node-by-node network thinking is poised to reach most nodes.

As networks slowly infect different areas of human inquiry, research on the fundamental properties of complex networks continues at full speed. Hardly a year has passed since *Linked* was published, but it would probably take another book to describe the array of discoveries that have surfaced during this period. Since it would be impossible to describe them all, I have set a more modest goal for the remainder of this afterword: I will revisit the impact of communities and modules on the structure of complex networks, an issue resolved only after *Linked* went to press.

1.

"Multitasking is counterproductive," claimed a CNN headline on December 6, 2002, questioning the increasingly popular business practice of forcing some employees to juggle a variety of tasks at a frequency that would challenge even a successful performer. Multitasking is a practice borrowed from one of our own creations, the computer: Humans mimic modern operating systems by performing several jobs simultaneously. Yet, as each job taxes memory and processor time, even computers struggle with too many tasks. The CNN coverage, based on a report published in the *Journal of Experimental Psychology*, indicated that humans and computers respond similarly to overloads—they slow down. Before turning to a new activity our brain must activate the skills and information required for the new task. A series of experiments instructing young adults to switch repeatedly between tasks of varying difficulty, such as solving a math problem or classifying geometric objects, indicated that in most cases the brain requires only a few seconds to perform this activity. The more complex the task ahead of us, the longer it takes the brain to take it on, with profound implications for businesses that see multitasking as the miracle antidote in a slow economy. Indeed, asking employers to frequently switch between diverse duties could reduce productivity as much as 20 to 40 percent instead of increasing it.

Despite these well-documented limitations, our brain and body are unparalleled in their ability to multitask. Indeed, as you read these lines, millions of visual receptors on your retina record changes in the intensity of the incoming light, pass this information to the visual cortex for processing, turning them into letters and words. Meanwhile, another network of cells guarantees that your heart continues to beat, and sophisticated neural circuits monitor a wide array of sensory inputs, from smells to temperature gauges. Sleep or awake, our brain and body juggle thousands of complicated tasks without signs of slowing down.

Our ability to multitask is inherited from our primitive ancestors. Indeed, even such simple organisms as the primitive *E. coli* bacteria run hundreds of cellular programs simultaneously, allowing them to swim toward food, take up molecules from their environment, and create offspring by duplication. Multitasking is an inherent property of most complex systems, alive or not. A company must purchase a wide array of raw materials from diverse vendors, run a manufacturing line, package and sell its products, lobby and carry on legal fights, develop new products, run ad campaigns in dozens of media, enter new markets on several continents with quite different business and legal environments. How can complex systems juggle thousands of tasks with such apparent ease? How does the need to multitask affect the architecture of a cell or a company? Answers to these questions are found under our network microscope as we consider an often overlooked property of complex systems: their modularity.

2.

Scientists can be notoriously shortsighted when it comes to their ability to predict the future. Indeed, in December 1899, barely three years before the Wright brothers achieved the first sustained powered flight, Maurice Levy, president of the *Paris Academie des Sciences* marked the turn of the twentieth century by voicing his opinion that heavier-than-air-flight would never succeed. Ernest Rutherford, a towering figure of atomic and nuclear physics, often called the potential usefulness of nuclear energy "all moonshine." Yet, only a year after Rutherford's death, in 1937, Otto Hann and Fritz Strassman demonstrated nuclear fission, prompting Albert Einstein to send his famous letter to President Roosevelt, opening the stage for the atomic bomb and nuclear energy.

Regardless of these much publicized failures of the scientific crystal ball, in 1999 *Nature* played tribute to the new millennium by asking several prominent scientists to peak into the twenty-first century. It is too early to judge how our generation measured up to this task. We do not have to wait a century, however, to recognize the impact of one of the papers commissioned by *Nature*. The cell biologist Leland H. Hartwell, recipient of the 2001 Nobel Prize in physiology or medicine for his work on cell division, joined forces with Andrew W. Murray and physicists John J. Hopfield and Stanislas Leibler to address a fundamental question of postgenomic biology: How are cells organized? Today the nearly century-old goal of characterizing the components of living systems is close to completion. Sophisticated databases contain inventories of most human and bacteria genes. Yet, this much awaited library failed to answer the question that has driven decades of research: What are the organizing principles behind life? How are the millions of cellular components integrated into a single living system? Hartwell and colleagues jumped into the middle of this debate by proposing that cells sustain a multitude of functions—i.e., multitask—thanks to a discrete modular organization. According to this view, the network behind the cell is fragmented into groups of diverse molecules, or modules, each module being responsible for a different cellular function. As modules are connected via a few links to other modules, the network behind the cell is similar to Granovetter's circles of friends, where those within the same circle know each other well and communication with other circles is maintained by a few weak ties. This modular hypothesis has deep roots in modern cell biology. For example, chemotaxis, the bacteria's ability to sense food and swim toward it, has been successfully reduced to a relatively autonomous functional module of a few key molecular components. Hartwell and colleagues went a step further by suggesting that chemotaxis is by no means an exception, but rather the norm: Distinct modules are responsible for most known cellular functions.

Modularity is a defining feature of most complex systems. Indeed, departmentalization allows large companies to create relatively secluded groups of employees who work together to solve specific tasks; the Web is fragmented into heavily interlinked communities of Webpages whose creators share common interests; modularity in our intellectual and professional interests allows Amazon to offer book recommendations inspired by the reading patterns of people within a comparable intellectual module; a modular computer design allows us to replace the old bulky screen with a flat panel display without redesigning the whole computer. Yet, a modular architecture is at odds with everything we have learned so far about complex networks. Most networks, from the cell to the World Wide Web, are scale free, held together by a few hubs. By virtue of the many links hubs possess, they must be in contact with nodes from numerous modules. Therefore modules cannot be that isolated after all, resulting in a fundamental conflict between the known scale-free architecture and the modular hypothesis. Current network models fail to resolve this contradiction: Neither the scale-free model nor the random network model of Erdős and Rényi shows signs of modularity. Yet, real networks are clearly scale free and appear to be modular at the same time, a paradox that fundamentally questions our understanding of how complex networks are organized.

3.

Two thousand and one was probably the busiest year of my life, dominated by two equally demanding projects. First, *Linked* included a daily writing schedule that typically started at 6 A.M. and ended late at night. Second, at the urging of Zoltán Oltvai, my biologist collaborator from Northwestern University, we tried to address the enduring mystery of small worlds—the fundamental conflict between hubs and modules. Writing a general audience book and focusing on science at the same time turned out to be quite a challenge. So the first hint toward recon-

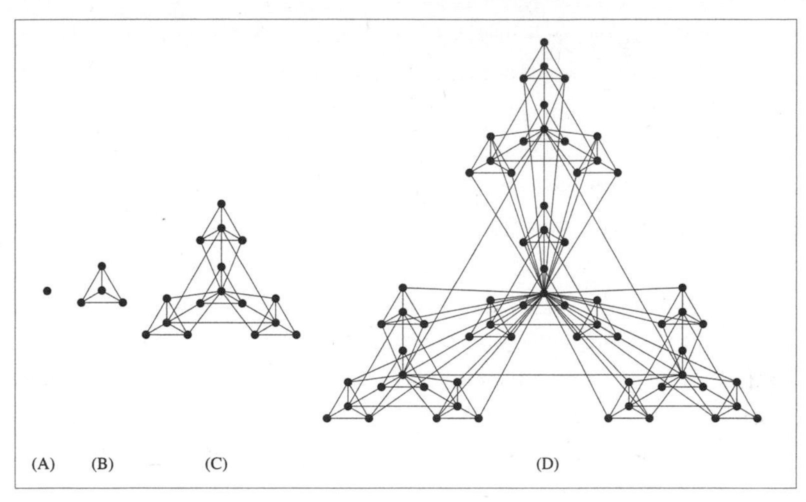

Figure 16.1 *We can generate a hierarchical network by starting from a single node (A) and making three copies of it, connecting the new nodes to the old node and to each other, obtaining the four-node structure shown in (B). In the next step we make three copies of our four-node structure and place them around the old module, connecting the peripheral nodes of the new modules to the central node of the old module and linking the central nodes of the new modules to each other. The obtained network will have sixteen nodes, as shown in (C). We can repeat the same process, creating now three copies of the C module and placing them around the old one, connecting the peripheral nodes to the center of the old module and the central in (D). The process can be continued indefinitely, each time generating a four-times-larger network. The obtained network is scale free: One can clearly distinguish a hierarchy of many small nodes, held together by a few large hubs. It is also modular, made of a hierarchy of larger and larger modules. Indeed, one can easily deconstruct the network shown in (C) into sixteen four-node modules, or four sixteen-node modules. An interesting property of the network is that it displays hierarchical clustering: It is made of many highly interlined four-node modules, which in turn form less interlinked sixteen-node modules, which are the building blocks of an even looser sixty-four-node module. Recently we learned that such hierarchical clustering is a generic property of a large number of real networks, from the cell to the World Wide Web.*

ciling hubs with modules came in the summer of 2001, when, together with Erzsébet Ravasz, a doctoral student in my group, and Tamás

Vicsek, my former thesis advisor from Eötvös University in Budapest, we constructed a little model that pointed toward a potential fusion between scale-free and modular networks.

To build a modular network we started with a single node (see Figure 16.1 A) and created three copies of it, connecting them to the old node and to each other, obtaining a little four-node module (B). We next generated three copies of this module, linking the peripheral nodes of each new copy to the central node of the old module, obtaining a sixteen-node network (C). Another "copy and link" step again quadrupled the number of nodes, resulting in a sixty-four-node network (D). While we could have continued this process indefinitely, we stopped here and inspected the intricate structure of the network. First, it was modular by construction: At the lowest organizational level it was made of many highly interconnected four-node modules. These modules were the building blocks of the larger sixteen-node modules, which in turn were the major components of the sixty-four-node network. Second, a highly connected central hub with thirty-nine links held the network together. The central nodes of the sixteen-node modules served as somewhat smaller local hubs, with fourteen links. Numerous nodes with a few links only accompanied these hubs, resulting in the familiar hierarchy of many small nodes held together by a few large hubs, a signature of scale-free networks. Indeed, the number of nodes with exactly *k* links followed a power law, confirming the model's scale-free nature[1] and successfully reconciling Granovetter's friendship circles with the hubs.

The scale-free model played an important role in our understanding of complex networks because it was the first to account for the power laws characterizing real networks. The new model offers the missing link: a modular scale-free network. How do we test, however, if it correctly reproduces the modularity of real systems? This was the question that we couldn't stop thinking about after coming up with the model. Failing to find a quick answer, we posted a short paper in July

1. For the construction described above, the degree distribution follows $P(k) \sim k^{-\eta}$ with $\eta = 1 + \ln 4/\ln 3 = 2.26$.

2002 on the Los Alamos preprint database with an early version of the model. Five months later we were still struggling with our question when Sergey Dorogovtsev, A. V. Goltsev, and José Mendes, working together in Porto, Portugal, posted on-line a short paper with the much awaited missing link.

A quantitative measure of modularity is the clustering coefficient (discussed in chapter 4), telling us how interlinked the neighbors of a node are. For a given node the clustering coefficient is equal to one if each of the node's neighbors are linked to each other. If the neighbors are not connected at all, the clustering coefficient is zero.[2] The discovery of hubs and power laws forced us to replace the democratic utopia of Erdős and Rényi, in which most nodes have the same number of links, with a scale-free world dominated by a few hubs.

When it comes to modules, we are about to experience a similar paradigm shift. Indeed, for both random and scale-free networks, small nodes and hubs have the same clustering coefficient, clustering being democratically distributed. The Porto trio arrived at a very different conclusion for the model discussed above: They showed that the more connected a node is, the smaller its clustering coefficient.[3] In particular, they found that modules follow a strict power law,[4] the clustering coefficient of a node with *k* links decreasing as the inverse of *k*. This represents a remarkable deviation from the predictions of the scale-free or random network models, for which the clustering coefficient is independent of *k*. Since the clustering coefficient is easy to measure, this new scaling law offered a powerful tool to explore the modularity of real networks.

---

2. The clustering coefficient of a node with k links is defined as $C(k) = 2N(k)/k(k-1)$, where $N(k)$ denotes the number of direct links between the node's k neighbors.
3. The central node of the four-node module (see figure B) has clustering coefficient equal to one, as each of its three neighbors has links to each other. The node at the center of the sixteen-node module (see figure C) is less clustered: Only twelve of its twelve neighbors have links to each other, thus $C = 2/14$. The node at the middle of the sixty-four-node construction has an even smaller C: Its thirty-nine neighbors are connected to each other by thirty-nine links (see figure D), giving $C = 2/38$.
4. The Porto group showed that the clustering coefficient $C(k)$ of a node with k links depends on k as $C(k) \sim k^{-1}$.

As our initial thinking about modularity was prompted by the desire to understand the inner organization of the cell, Erzsébet Ravasz, Zoltán Oltvai, and I first investigated the metabolism of forty-three organisms. The results were unambiguous: For each organism the clustering coefficient followed the power law predicted by the model. Therefore, the cell is not only modular, but its modularity has a strict architecture: Numerous small but highly interlinked modules combine in a hierarchical fashion into a few larger, less interlinked modules. There are no "typical" or "characteristic" modules in the cell. Rather, the metabolism can be equally well deconstructed into many small, highly interlinked modules or into a few larger but less cohesive ones. Hierarchical modularity sheds new light on the role of the hubs as well: They maintain communication between the modules. Small hubs have links to nodes belonging to a few smaller modules. Large hubs act like the CEO of a major company, having jurisdiction over most departments and modules, bridging together communities of different sizes and cultures.

Hierarchical modularity has significant design advantages: It permits parts of the system to evolve separately. Indeed, car manufacturers can improve the transmission of a car without rebuilding the engine and chip designers can build faster processors without worrying about the keyboard or the CD-ROM. In natural systems, modularity allows evolution to experiment separately with individual functions. Indeed, the impact of genetic mutations, affecting at most a few genes at once, is limited to a few modules. If a mutation is an improvement, the organism with the superior module will flourish. If, however, tinkering with a gene decreases the module's fitness, the organism will fail to survive.

The subtle hierarchical modularity discussed above is by no means limited to the cell. Indeed, rereading the recent literature on networks we realized that quantitative signatures of modularity have been spotted before. Jean-Pierre Eckmann from the University of Geneva in Switzerland with Elisha Moses from the Weizmann Institute of Science in Israel observed that on the World Wide Web the clustering coefficient follows a power law, adding that, "while the scaling is striking, we are not sure of its origin." In hindsight, their finding offers direct proof of the Web's hierarchical architecture. For example, the Kurt Vonnegut or Philip Roth small community of fanpages are nested in a hierarchical fashion into the larger group of Webpages focusing on American literature. Similarly, Alexei Vazquez, Romualdo Pastor-Satorras, and AlessandroVespignani have found that domain-level Internet maps have a hierarchical organization, marked by the scaling of the clustering coefficient. Our measurements indicate that language, viewed as a network of synonyms, is hierarchical as well, a few highly connected words like "turn," "take," or "go," each with over a hundred synonyms, holding the various lexical modules together. The hierarchical modularity of the protein interaction network offered another biological example, reinterpreting the role of the hub proteins as the mediators of different functional modules. Therefore, hierarchical modularity is a generic property of most real networks, accompanying the scale-free architecture.

It is this hierarchical modularity that makes multitasking possible: While the dense interconnections within each module help the efficient accomplishment of specific tasks, the hubs coordinate the communication between the many parallel functions. Bottlenecks and slowdowns are inevitable if the same module is simultaneously confronted with several tasks. The computer's dependence on a single central processing unit is its main bottleneck, and when our cerebral cortex is taxed with too many tasks, we slow down too.

4.

While noticing similarities between natural systems and human designs has a long history, it has always been difficult to move beyond the metaphors to quantify these analogies. Lately, networks have become the X-ray machines of our connectedness, diagnosing the cell or the Web with the same ease. Thanks to the rapid advances in network theory it appears that we are not far from the next major step: constructing a general theory of complexity. The pressure is enormous. In the twenty-first century, complexity is not a vague science buzzword any longer, but an equally pressing challenge for everything from the economy to cell biology. Yet, most earlier attempts to construct a theory of complexity have overlooked the deep link between it and networks. In most systems, complexity starts where networks turn nontrivial. No matter how puzzled we are by the behavior of an electron or an atom, we rarely call it complex, as quantum mechanics offers us the tools to describe them with remarkable accuracy. The demystification of crystals—highly regular networks of atoms and molecules—is one of the major success stories of twentieth-century physics, resulting in the development of the transistor and the discovery of superconductivity. Yet, we continue to struggle with systems for which the interaction map between the components is less ordered and rigid, hoping to give self-organization a chance.

The mystery of life begins with the intricate web of interactions, integrating the millions of molecules within each organism. The enigma of the society starts with the convoluted structure of the social network. The unpredictability of economic processes is rooted in the unknown interaction map behind the mythical market. Therefore, networks are the prerequisite for describing any complex system, indicating that complexity theory must inevitably stand on the shoulders of network theory. It is tempting to step in the footsteps of some of my predecessors and predict whether and when we will tame complexity. If nothing else, such a prediction could serve as a benchmark to be disproven. Looking back at the speed with which we disentangled the networks around us after the discovery of scale-free networks, one thing is sure: Once we stumble across the right vision of complexity, it will take little to bring it to fruition. *When* that will happen is one of the mysteries that keeps many of us going.

Acknowledgments

IT IS IMPOSSIBLE TO PROPERLY ACKNOWLEDGE all those scientists whose research has inspired me. Some are mentioned in the book; others appear in the notes. The vast majority, however, did not find their way into the narrative. But they are no less important. In many ways, they form the foundation of this book.

My sources were not limited to published materials. I greatly benefited from all those students and colleagues who have approached me over the years with questions and ideas about networks. These conversations and emails have played a crucial role in shaping the book's content, opening my eyes to network thinking in many different areas.

This book would have never been possible without the help of my truly extraordinary former and current students, Réka Albert, Ginestra Bianconi, Zoltán Dezső, Illés Farkas, Erzsébet Ravasz, Soon-Hyung Yook, who chose to research networks. The exceptional skills and perseverance of Hawoong Jeong, a former postdoc in my research group, made our journey towards real networks possible and fun. I am also indebted to numerous collaborators, including Eric Berlow, Jay Brockman, Imre Derényi, Jennifer Dunne, Vincent Free, Byungnam Kahng, János Kertész, Neo Martinez, Zoltán Néda, Zoltán Oltvai, Peter Schiffer, András Schubert, Bálint Tombor, Yuhai Tu, Tamás Vicsek, and Rich Williams, for the many ideas and help offered during the last few years.

Several people provided information that turned out to be crucial when collecting material for this book. Back in 1999, Tibor Braun mailed me the Karinthy short story about five degrees and Thomas Bass helped me understand Milgram's work on six degrees. Luis Amaral, Eric Bonabeau, Guido Caldarelli, Christopher R. Edling, Lee Giles, Mark Granovetter, Byungnam Kahng, Jeff Kantor, Judith Kleinfeld, Valdis Krebs, Steve Lawrence, Fredrick Liljeros, Sid Redner, Ricard V. Solé, and Alessandro Vespignani shared unpublished manuscripts, information and sources with me.

Many people have volunteered their time to read and comment on the manuscript in its various stages. Zoltán Oltvai has done a truly heroic job, offering line-by-line suggestions to improve it. I have also greatly benefited from the comments and insights of Réka Albert, Kevin Barry, Steve Buechler, Reuven Cohen, Gábor Forgács, Viktor Gyuris, Boldizsár Jankó, Hawoong Jeong, Gerald Jones, Jim McAdams, Mark Newman, H. Eugene Stanley, Alessandro Vespignani, Tamás Vicsek, Ed Vielmetti and Eduardo Zambrando. The truly stimulating lunch discussions with members of the Web Group, a group of Notre Dame faculty steered by Jim Mcadams, have helped me to refine some of the ideas that eventually made it into the book. The group included Sheri Alpert, Kevin Barry, Patricia Louise Bellia, Kathy Biddick, Jay Brockman, Leo Burke, David S. Hachen, Martin Haenggi, Lionel M. Jensen, Lee Byung Joo, Jeffrey C. Kantor, Barry Patrick Keating, Gregory Madey, Gail Hinchion Mancini, Khalil Matta, Kajal Mukhopadhyay, Daniel Myers, Susan Ohmer, and Richard Pierce.

My colleagues in the Physics Department, who have encouraged the wide turns my work has taken in the last few years, have created a truly supportive atmosphere for research and for writing this book. I wish to thank in particular Bruce Bunker, Margaret Dobrowolska, Jacek Furdyna, Boldizsár Jankó, Gerald Jones, and Kathie Newman for their continued support. Jennifer Maddox, with her professional and upbeat attitude toward every logistical problem I could dream up, saved me countless hours, all of which were poured into this book.

Deborah Justice spent many hours editing the early stages of the manuscript, helping me to understand the intricacies of English grammar and style. The book would have never been finished without the dedicated technical help of Lesley Krueger, who witnessed first hand the metamorphosis of a rough draft into a ready book. Enikő Jankó and Zoltán Dezső spent many hours updating the never-ending versions of the manuscript, jumping in at short notice and at odd times. I am truly indebted to Hawoong Jeong, who time and again displayed his magic touch and infinite patience in preparing the final figures for the book.

The staff at Perseus has been extremely supportive and smart at every step of the writing and production process. I cannot imagine better editors than Amanda Cook and Joan Benham, whose editorial guidance made the book far more readable. Elizabeth Carduff, Chris Coffin and Marco Pavia guided the manuscript through the labyrinth of the production process with exceptional speed. Many thanks to Lissa Warren, who helped get the word out. I am grateful to Katinka Matson and Brockman, Inc. for pairing me up with Perseus.

Linked could have never been started—let alone finished—without the gentle enthusiasm of Janet Kelley, who encouraged me each time I was ready to give up. It is hard to put into words how much her support has meant to me.

Notes

THE FIRST LINK: INTRODUCTION

Page 2 The story of MafiaBoy has been widely discussed in the press. For a collection of links, see http://www.mafiaboy.com/. The "Yes, I heard you" phrase comes from C. Taylor, "Behind the Hack Attack," *Time Magazine* (February 21, 2000).

Page 3 St. Paul's life and role in spreading Christianity have been the focus of countless books and monographs. See, for example, C. J. Den Heyer, *Paul: A Man of Two Worlds* (Harrisburg, Penn.: Trinity Press International, 1998) and Robert Jewlett, *A Chronology of Paul's Life* (Philadelphia: Fortress Press, 1979).

Page 6 Complexity, an emerging field of science aiming to understand how systems made of millions of diverse components behave and how order emerges from chaos and randomness, riding the laws of self-organization, is booming lately. Its practitioners span dozens of disciplines, ranging from mathematics and physics to ecology and business. For popular books and introductory texts on the subject, see for example Murray Gell-Mann's *The Quark and the Jaguar: Adventures in the Simple and the Complex* (New York: W. H. Freeman, 1995); *Hidden Order: How Adaptation Builds Complexity* (Cambridge, Mass.: Perseus, 1996); Ricard V. Solé and Brian Goodwin's *Signs of Life: How Complexity Pervades Biology* (New York: Basic Books, 2001); Yaneer Bar-Yam, *Dynamics of Complex Systems* (Cambridge, Mass.: Perseus, 1997).

THE SECOND LINK: THE RANDOM UNIVERSE

Page 9 Euler's life has been briefly described in many different places. For a recent account, see for example the biographical note in Willam Durham, *Euler: The Master of Us All* (Washington, D.C.: Mathematical Association of America, 1999).

Page 10 Euler's life in St. Petersburg was rather tumultuous. He lost his wife of three decades, only to marry his wife's half sister three years later. A fire burned down his house and destroyed all his books and notes; he escaped thanks to the bravery of a fellow Swiss who walked into the fire and carried his illustrious countryman out on his back.

Page 10 The freshness and clarity of Euler's writing is remarkable. His enthusiasm for his subject vibrates even centuries later. Today, when scholarly writing is hardly intelligible to the uninitiated, the simple language of his papers is enviable. Euler spent about a quarter century in Berlin, invited by Frederick the Great, King of Prussia, during which time he was asked to offer lessons in the natural sciences to the Princess of Anhalt Dessau, the King's niece. The four-hundred-page work written as a response to this assignment, *Letters of Euler on Different Subjects in Natural Philosophy, Addressed to a German Princess*, touches on just about all fields of science, discussing issues ranging from the gravity of the Moon to the nature of spirits. *Letters to a German Princess* soon became an international hit. This is the finest example of popular science writing by a scholar working at the frontiers of his field. See Leonhard Euler, *Letters of Euler on Different Subjects in Natural Philosophy Addressed to a German Princess* (New York: Arno Press, 1975).

Page 10 Euler's collected works are published under the title *Opera Omnia* (Basel, Switzerland: Birkhäuser Verlag AG, 1913). Though their publication began in 1911 and over six dozen volumes have appeared, the work is still incomplete.

Page 10 A nice account of the early days of graph theory, with historical background and English translation of the most important papers in the field, can be found *in Graph Theory: 1736–1936*, by Norman L. Biggs, E. Keith Lloyd, and Robin J. Wilson (Oxford, England: Clanderon Press, 1976). The book also contains Euler's original paper on the Königsberg bridges.

Page 12 To see in more detail Euler's arguments, consider, for example, node *D*, which has three links representing land connected to three bridges, *e*, *f*, and *g*. A person attempting to cross all three bridges only once has to visit node D at least twice. For example, he could arrive through bridge *f*, leave on *e*, and then come again on *g*. The problem is that he cannot leave again, since there is no unvisited bridge left. Thus *D* is either the starting point or the end point of the journey. But this is not a unique feature of node *D*: It is easy to check that *all nodes* with an *odd* number of links have this property. That is, a person who wants to visit all nodes must start or end on them. Nodes *A*, *B*, C and *D* share this property, too, as they all have three links each. This means that A, *B*, C, and *D* have to be simultaneously the starting or ending point of the journey.

Page 13 For example, the four color problem, dating back to 1852, was missing a proof until 1976. The problem is very simple at the outset: Prove that any map can be colored with four colors such that no two countries sharing a border have the same color. Anyone who has tried to color a map can easily convince him or herself that, indeed, four colors are sufficient. Yet the proof evaded mathematicians for well over a century, becoming the first major theorem to have a computer-aided proof.

Page 14 Erdős's life has inspired many popular accounts. The story in the introduction is based on the account of András Várzsonyi, who was the fourteen-year-old in the shoe store and who later became the second-youngest math Ph.D. in Hungary in that era (after Erdős) and Erdős's lifelong friend. See Fan Clung and Ron Graham, *Erdős on Graphs: His Legacy and Unsolved Problems* (Wellesley, Mass.: A. K. Peters, 1998). For popular Erdős biographies see Paul Hoffman, *The Man Who Loved Only Numbers* (New York: Hyperion, 1998) and Bruce Schechter, *My Brain Is Open* (New York: Touchstone, 1998). See also András Hajnal and Vera T. Sós, "Paul Erdős Is Seventy," *Journal of Graph Theory* 7 (1983): 391–393.

Page 14 The honor roll of eight articles by Erdős and Rényi, that established random graph theory is listed in Michal Karonski and Adrzej Rucinski, "The Origins of the Theory of Random Graphs," in *The Mathematics of Paul Erdős*, ed. R. L. Graham and J. Nesetril (Berlin: Springer, 1997). They are *On Random Graphs I*, Math. Debrecen vol. 6, 290–297 (1959). "On the Evolution of Random Graphs," *Publ. Math. Inst. Hung. Acad. Sci* 5 (1960): 17–61; "On the Evolution of Random Graphs," *Bull. Inst. Internat. Statist* 38 (1961): 343–347. "On the Strength of Connectedness of a Random Graph," *Acta Math. Acad. Sci. Hungar* 12 (1961): 261–267. "Asymmetric Graphs," *Acta Math. Acad. Sci. Hungary* 14 (1963): 295–315; "On Random Matrices," *Publ. Math. Inst. Hung. Acad. Sci* 8 (1964): 455–461; "On the Existence of a Factor of Degree One of a Connected Random Graph," *Acta Math. Acad. Sci. Hungary* 17 (1966): 359–368; "On Random Matrices II," *Studia Sci. Math. Hung* 13 (1968): 459–464.

Page 17 Note that unknown to Erdős and Renyi, and to *most* mathematicians, random networks had been introduced a decade before the classic Erdős-Rényi work, in Ray Solomonoff and Anatol Rapoport, "Connectivity of Random Nets," *Bulletin of Mathematical Biophysics*, 13 (1951): 107–227. Interestingly, the paper derives the classical result again attributed to Erdős and Rényi, that when the average degree reaches one, a giant cluster emerges. It is difficult to explain why the paper was never recognized as a precursor to Erdős and Rényi's work. Perhaps the mathematical beauty of Erdős's proof was the appealing aspect for mathematicians, something that the heuristic derivation of Solomonoff and Rapoport was clearly lacking.

Page 20 Rényi's life, however, is much less covered in the international press. For a tour of his mathematics and life, see the series of articles published soon after his death (in Hungarian) in *Matematikai Lapok* 3–4 (1970): Turán Pál, "Rényi Alfréd munkássága," 199–210; Révész Pál, "Rényi Alfréd valószinüségszámitási munkássága," 211–231; Csiszár Imre, "Rényi Alfréd információelméleti munkássága," 233–241; Katona Gyula and Tusnády Gábor, "Rényi Alfréd pedagógiai munkássága, 243–244; B. Mészáros Vilma, "Guibus Vivere est Cogitare," 245–248. Today he is remembered through the recently renamed hotbed of Hungarian mathematics, the Alfréd Rényi Institute of Mathematics in Budapest.

Page 20 Though he failed to attract Erdős to the Notre Dame faculty, Arnold Ross went on to a very prolific career as a scientific educator. He founded the Ross Program, an intensive summer course on mathematics for talented high school students and teachers. Ross started his program in 1947 at Notre Dame and moved it to Ohio State University in 1964, and has run it every summer since. See, e.g., Allyn Jackson, "Interview with Arnold Ross," *Notices of the American Mathematics Society* 48, no. 7 (August 2001): 691–697.

Page 22 The degree distribution of a random graph has been derived in B. Bollobás, "Degree Sequences of Random Graphs," *Discrete Mathematics* volume 33, pg. 1 (1981).

Page 23 In retrospect it is not clear how much the work of Erdős and Rényi was motivated by a desire to model our interlinked world and how much by the mathematical beauty of the problem at hand. In their seminal 1959 paper they do mention potential applications: "It seems plausible that by considering the random growth of more complicated structures . . . one could obtain fairly reasonable models of more complex real growth processes (e.g. the growth of a complex communication net consisting of different types of connections, and even of organic structures of living matter, etc.)." Despite this incredible insight into the future, it is fair to assume that their work in this area was motivated by a deep curiosity about the mathematical depths of the problem rather than by its applications.

Page 23 As we continue our journey into our interconnected world, the Erdős-Rényi theory of random networks will often be our reference point. Moving along the links, we cannot help occasionally contrasting it with the real world. Yet, the shortcomings of the model do not decrease our admiration and respect for Erdős and Rényi's monumental legacy. Our occasional criticism is aimed at all of us who, as their followers, for decades indiscriminately applied to the real world the random worldview they depicted.

THE THIRD LINK: SIX DEGREES OF SEPARATION

Page 25 There are a large number of books in Hungarian about Karinthy's work and life. See, e.g., the *in memoriam* volume Karinthy Frigyes, *A humor a teljes igazság*, ed. Mátyás Domokos, Budapest: Nap Kiadó, 1998), which is a collection of stories about Karinthy written by friends and colleagues, most of them writers. See also Dolinszky Miklós, *Szószerint (A Karinthy Passió)*, (Budapest: Magvető, 2001), and Levendel Júlia, *Így élt Karinthy Frigyes* (Budapest: Móra Könyvkiadó, 1979).

Page 26 Frigyes Karinthy, "Láncszemek," in *Minden másképpen van* (Budapest: Atheneum Irodai es Nyomdai R.-T. Kiadása, 1929), 85–90. I am deeply indebted to Tibor Braun, who brought this story to my attention and mailed me a copy in 1999, soon after our research on nineteen degrees on the World Wide Web was featured in the Hungarian media. I do not think that the short story has ever been translated into English. For a collection of short stories in English by Karinthy, see Frigyes Karinthy, *Grave and Gay* (Budapest: Korvina Kiadó, 1973).

Page 27 For another early appearance of the six-degrees concept see Jane Jacobs's *The Death and Life of American Cities* (New York: Random House, 1961), one of the most important books ever written about city planning, setting off the revival of old-fashioned neighborhoods. In this book she recalls, "When my sister and I first came to New York from a small city, we used to amuse ourselves with a game we called Messages. The idea was to pick two wildly dissimilar individuals—say a head hunter in the Solomon Islands and a cobbler in Rock Island, Illinois—and assume that one had to get a message to the other by word of mouth; then we would each silently figure out a plausible, or at least possible, chain of persons through which the message could go. The one who could make the shortest plausible chain of messengers won."

Page 27 Milgram's six-degrees study was published in several places. See for example Stanley Milgram, "The Small World Problem," *Physiology Today* 2 (1967): 60–67. For his work on obedience, which is fascinating reading by itself, see Stanley Milgram, *From Obedience to Authority* (New York: Harper and Row, 1969).

Page 27 To learn more about the work and life of Milgram, see Thomas Blass, "The Social Psychology of Stanley Milgram," in *Advances in Experimental Social Psychology*, ed. M. P. Zanna (San Diego: Academic Press, 1992), 25:277–328; and Thomas Blass, ed., *Obedience to Authority: Current Perspectives on the Milgram Paradigm* (Mahwah, N.J.: Lawrence Erlbaum, 2000). Further information and links can be obtained from http://www.stanley milgram.com.

Page 28 Note that the methodology used by Milgram to reach the six-degrees conclusion has been questioned recently by Judith S. Kleinfeld, who inspected Milgram's papers and notes as well as the collection of letters that made their way to their destination, kept in the Yale Archives. In particular, recent studies suggest that our world is deeply divided by class and race, making navigation across such social barriers rather difficult. See, e.g., Judith S. Kleinfeld, "The Small World Problem," *Society* 39 (January-February, 2002): 61–66; and "Six Degrees of Separation: An Urban Myth," *Psychology Today* (forthcoming in 2002).

Page 29 The technical aspects of Milgram's work on our social connectivity was known to have been inspired by the work of Anator Rapaport, a Russian-born mathematician and concert pianist who published several seminal papers on social networks. He independently introduced the concept of random graphs and had a huge impact on sociology. His most relevant papers include R. Solomonoff and A. Rapaport, "Connectivity of Random Nets," *Bulletin of Mathematical Biophysics* 13 (1951): 107–117; and A. Rapaport, "Contribution to the Theory of Random and Biased Nets," *Bulletin of Mathematical Biophysics* 19 (1957): 257–277.

Page 29 John Guare, *Six Degrees of Separation* (New York: Random House, 1990).

Page 30 For the early history of the World Wide Web as told by the Web's creator, see Tim Berners-Lee with Mark Fischetti, *Weaving the Web: The Original Design and Ultimate Destiny of the World Wide Web by Its Inventor* (San Francisco: Harper, 1999).

Page 31 Regarding the size of the World Wide Web, see Steve Lawrence and C. Lee Giles, "Searching the World Wide Web," *Science* 280 (1998): 98–100; and "Accessibility of Information on the Web," *Nature* 400 (1999): 107–109. See also the extensive discussion in Chapter 12.

Page 34 The study uncovering the nineteen degrees of the World Wide Web was published in R. Albert, H. Jeong, and A.-L. Barabási, "Diameter of the World Wide Web," *Nature* 401 (1999): 130–131. The method we used to uncover the diameter of the full Web is known in the literature as finite size scaling.

Page 34 For the degree of separation in food webs, see Richard J. Williams, Neo D. Martinez, Eric L. Berlow, Jennifer A. Dunne, and Albert-László Barabási, *Two Degrees of Separation in Complex Food Webs*, http://www.santafe.edu/sfi/publicatons/Abstracts/01-07-036 abs.html; José M. Montoya and Ricard V. Solé, *Small World Patterns in Food Webs*, http://www.santafe.edu/sfi/publications/Abstracts/00-10-059abs.html.

For the separation within the cell, see Chapter 13 and Hawoong Jeong, Bálint Tombor, Réka Albert, Zoltán N. Oltvai, and Albert-László Barabási, "The Large-Scale Organization of Metabolic Networks, *Nature* 407 (2000): 651; Hawoong Jeong, Sean Mason, Albert-László Barabási, and Zoltán N. Oltvai, "Centrality and Lethality of Protein Networks," *Nature* 411 (2001): 41–42; Andreas Wagner and David Fell, *The Small World Inside Large Metabolic Networks*, Proceedings of the Royal Society of London, Series B—Biological Sciences, vol. 268 (Sept. 7, 2001): 1803-1810. For the network of scientists and the small world between them, see A.-L. Barabási, H. Jeong, E. Ravasz, Z. Néda, T. Vicsek, and A. Schubert, *Evolution of the Social Network of Scientific Collaborations*, http:///xxx.lanl.gov/abs/cond-mat/0104162 (forthcoming in 2002); M. E. J. Newman, *Who is the Best Connected Scientist? A Study of Scientific Coauthorship Networks*, http://www.santafe.edu/sfi/publications/Abstracts/00-12-064abs.html; M. E. J. Newman, *The Structure of Scientific Collaboration Networks*, Proceedings of the National

Academy of Sciences of the United States of America, vol. 98, (Jan. 16, 2001): 404–409. For the small world in neural networks see D. J. Watts and S. H. Strogatz, "Collective Dynamics of 'Small-World' Networks," *Nature* 393 (1998): 440–442.

Page 35 We can test our simple prediction on two familiar networks: society and the Web. For society we need to know how many people an average person knows. Though this might seem simple to calculate, sociologists cannot agree, and estimates range from 200 to over 5,000! As Duncan Watts, a mathematician turned sociologist at Columbia University, told me recently, finding the right answer is complicated by the difficulty of defining the term *acquaintance*: "I probably know the names of a few thousand people, but would I call them when visiting their city? Would I feel comfortable asking them for a favor? Would I confide in them?" To get around this problem, let's assume that an average person has about 1,000 acquaintances whom he or she knows on a first-name basis, which is about halfway between the most conservative and the most optimistic estimates. With 6 billion people, our formula tells us that the separation in society is close to three. The same formula applied to the World Wide Web, using a billion documents and an average connectivity of seven, predicts a separation of ten. Both predictions are on the lower end but not that far from the "correct" answers (six and nineteen, respectively). The mathematical formula washes away all but the order-of-magnitude differences in the number of nodes, explaining why we get such a small separation in general.

The deviation from the "correct" answer underlies the basic premise of this book: Real networks are not random. If the World Wide Web were random, since there is little ambiguity about its connectivity and size, its separation would be much closer to the prediction of ten clicks of the random network formula. That would be true for many other networks in nature, as well, for which size and connectivity are known rather precisely. The numbers, however, rarely match—a hint of the order hidden within our interwoven world. For a detailed list of small-world networks, highlighting the discrepancy between the random network prediction and their real separation, see, e.g., R. Albert and A.-L. Barabási, "Statistical Mechanics of Complex Networks," *Reviews of Modern Physics* 74 (January 2002): 47–97.

Page 36 For Kochen's note on the history of small worlds see Manfred Kochen, preface to *The Small World*, ed. Manfred Kochen (Norwood, N.J.: Ablex, 1989).

Page 38 For a discussion on the navigability of small worlds, see J. M. Kleinberg, "Navigation in a Small World—It Is Easier to Find Short Chains Between Points in Some Networks Than Others," *Nature* 406 (August 2000): 845.

Page 39 Note that one can also argue that Milgram's work has underestimated the degree of separation between people, as it did not correct for the effect of incomplete chains. Indeed, if a letter did not make it to the target person, it was ignored. As in the Nebraska experiment, only 42 of 160 letters arrived to the target, the majority of larger chains was clearly absent from the sample. Longer chains have a larger probability of not making it, thus the sample studied by Milgram was biased toward the shorter completed paths.

The Fourth Link: Small Worlds

Page 42 The original finding about our clustered society was published in Mark S. Granovetter, "The Strength of Weak Ties," *American Journal of Sociology* 78, (1973) 1360–1380. Granovetter recalled the saga surrounding its publication in the short article published in *Current Contents* (Sociology and Behavioral Sciences Edition, vol. 18, no. 49 [Dec. 1986]: 24) on the occasion of the paper being named a Citation Classic. See also "The Strength of Weak Ties: A Network Theory Revisited," *Sociological Theory* 1 (1983): 201–233; Mark S. Granovetter, *Getting a Job* (Cambridge, Mass.: Harvard University Press, 1994).

Page 44 Rhythmic applause has been extensively studied in the physics literature as a manifestation of synchronization. The first detailed measurements were reported in Z. Néda, E. Ravasz, Y. Brechet, T. Vicsek, and A.-L. Barabási, "Self-Organizing Processes: The Sound of Many Hands Clapping," *Nature* 403 (2000): 849–850 . For a more detailed account see Z. Néda, E. Ravasz, T. Vicsek, Y. Brechet, and A.-L. Barabási, "Physics of the Rhythmic Applause," *Physical Review* E 61, no. 6 (2000): 6987–6992. For a popular account of this research see Henry Fountain, "Making Order Out of Chaos When a Crowd Goes Wild," *New York Times*, March 7, 2000; and Josie Glausiusz, "Joining Hands," *Discover* 21 (July 2000).

Page 45 John Buck and Elisabeth Buck, "Synchronous Fireflies," *Scientific American*, May 1976, 74–85. For a recent book on synchronization see Arkady Pikovsky, Michael Rosenblum, and J. Kurths, *Synchronization: A Universal Concept in Nonlinear Sciences* (Cambridge, England: Cambridge University Press, 2001). See also Ian Stewart and Steven H. Strogatz, "Coupled Oscillators and Biological Synchronization," *Scientific American*, Dec. 1993, 68.

Page 46 For the story behind the Watts-Strogatz discovery, see Duncan J. Watts, *Small Worlds* (Princeton, N.J.: Princeton University Press, 1999).

Page 46 While networks diverted Duncan Watts' attention from synchronisation, a number of researchers have revisited the link between networks and synchronisation. See for example J. Jast and N. P. Jog, "Spectral Properties and Synchronisation in Coupled Map Lattices," *Physical Review*, E 65 (2002): 016201; X.F. Wang and G.R. Chen, "Synchronisation in a Scale-Free Dynamical Network: Robustness and Fragility, "IEEE Transaction on Circuits and Sysytems I 49 (2002):54–62; M. Barahona and L. M. Pecora, "Synchronisation in Small-World Systems," http://xxx.lanl.gov/abs/nlin.CD/0112023; J. Ito and K. Kaneko, "Spontaneous Structure Formation in a Network of Chaotic Units with Variable Connection Strengths," *Physical Review Letters*, 88 (2002): 02801.

Page 46 The term *clustering coefficient* was first used by Watts and Strogatz in D. J. Watts and S. H. Strogatz, "Collective Dynamics of 'Small-World' Networks," *Nature* 393 (1998): 440–442. The same quantity, under the name of "fraction of transitive triplets," was used in the sociology literature. See, e.g., the now classic S. Wasserman and K. Faust, *Social Network Analysis: Methods and Applications* (Cambridge, England: Cambridge University Press, 1994), 598–602.

Page 47 For an extensive discussion on Erdős numbers, see the Erdős site maintained by Jerrold W. Grossman, http://www.oakland.edu/~grossman.erdoshp.html. For a list of famous scientists and their Erdős number see Rodrigo De Castron and Jerrold W. Grossman, "Famous trails to Paul Erdős," *Mathematical Intelligencer*, 21 (Summer 1999): 51–63.

Page 48 Note that in particle physics, which is a very active subfield of physics, it is common that hundreds of physicists scattered over many continents and often never introduced to each other contribute to the discovery of a new elementary particle. Thus for them coauthorship cannot be taken as a signature of acquaintanceship or social ties. In most research fields, however, such large collaborations are the exception rather the norm.

Page 49 Our work on collaboration networks between mathematicians and neuroscientists is summarized in A.-L. Barabási, H. Jeong, R. Ravasz, Z. Néda, T. Vicsek, and A. Schubert, *On the Topology of Scientific Collaboration Networks*, http://xxx.lanl.gov/abs/ cond-mat/0104162 (forthcoming in *Physica* A, 2002). Similar results were obtained independently by Mark Newman, focusing on physicists, computer scientists, and other fields. See M. E. J. Newman, "The Structure of Scientific Collaboration Networks," *Proceedings of the National Academy of Sciences* 98 (2001): 404–409; "Scientific Collaboration Networks: I. Network Construction and Fundamental Results," *Physical Review*, E 64 (2001): 016131; "Scientific Collaboration Networks: II. Shortest Paths, Weighted Networks, and Centrality," *Physical Review*, E 64 (2001): 016132.

Page 49 What would the random network model tell us about the clustering coefficient? Since the clustering coefficient is the probability that two of my friends are connected, for the Erdős-Rényi model that is nothing but the probability that *any* two nodes have a link between them. Indeed, we can calculate the clustering coefficient for a random graph with N nodes by dividing the total number of links that are present in the network (L) with the total number of possible links $N(N-1)/2$, giving $C = 2L/N(N - 1)$, which turns out to be the control parameter of the Erdős-Rényi model, often denoted by p, denoting the probability that any two nodes are connected to each other. In other terms, the clustering coefficient is given by $C = \langle k \rangle / N$, where $\langle k \rangle$ is the average number of links per node in the network.

Page 49 With a bit of familiarity about how scientists collaborate, we can begin to understand the origin of the enormous clustering uncovered by Newman's and our study. Indeed, many scientific papers are written by three or more authors. Each such paper generates a complete graph, similar to our circle of friends, because each author is connected to all other authors of the paper. Thus the collaboration graph is sprinkled with tiny complete graphs, each of which has a high clustering coefficient, increasing the average clustering of the whole network. But there is a social factor too. Even if two of my graduate students did not coauthor a paper while in graduate school, each of them is still only two steps from each other, since they all wrote papers with me. Continuing to work in the same field, they will likely sometime in the future write a paper together, enhancing my clustering coefficient.

Page 50 For further information about the worm *Caenorhabditis elegans* see http://elegans. swmed.edu/ or http://www.nematodes.org/.

Page 50 While the C. *elegans* worm's neural wiring diagram has been fully mapped out, it would be impossible to attempt something similar for the human brain. This is not because there are billions of neurons to map out, some with thousands of links to other neurons, but rather because the connections between neurons in the human brain change, creating new links and severing old ones as we learn and age. C. *elegans*, however, stands out as a reliable study with a statically wired brain, in which connections are genetically predetermined.

Page 50 For a study on the topology of the *Caenorhabditis elegans* wiring diagram, see D. J. Watts and S. H. Strogatz, "Collective Dynamics of 'Small-World' Networks," *Nature* 393 (1998): 440–442, which contains the study on the power network and Hollywood actors, as well. See also S. Horita, K. Oshio, Y. Osama, Y. Funabashi, K. Oka, K. Kawamara, "Geometrical Structure of the Neuronal Network of Caenorhabditis Elegans," *Physica* A, 298 (2001): 553–561.

Page 50 For clustering in the World Wide Web, see L. A. Adamic, "The Small World Web," *Proceedings of the European Conference on Digital Libraries 1999 Conference* (Berlin: Springer Verlag, 1999): 443. For clustering in the Internet topology see, e.g., Soon-Hyung Yook, Hawoong Jeong, Albert-László Barabási, *Modeling the Internet's Large-Scale Topology*, http://xxx.lanl.gov/abs/cond-mat/0107417; and Romualdo Pastor-Satorras, Alexei Vazquez, Alessandro Vespignani, *Dynamical and Correlation Properties of the Internet, Physical Review Letters*, 2001: Article no. 258701. Clustering in the economy is discussed in Bruce Kogut and Gordon Walker, "The Small World of Germany and the Durability of National Networks," *American Sociological Review* 66 (2001): 317–335. For clustering in the ecological network see Richard J. Williams, Neo D. Martinez, Eric L. Berlow, Jennifer A. Dunne, and Albert-László Barabási, *Two Degrees of Separation in Complex Food Webs*, http://www.santafe.edu/sfi/publications/Abstracts/01-07-036abs.html.

Page 51 For additional examples of clustering in complex networks, see R. Albert and A-L. Barabási, "Statistical Mechanics of Complex Networks," *Reviews of Modern Physics* 74, No. 1 (January 2002), 47-97.

Page 52 The original model suggested by Watts and Strogatz in *Nature* 393 (1998): 440–442, did not add extra links but rewired some of the existing links to distant nodes, with identical effect. The version described here was proposed by M. Newman and D. J. Watts; see, e.g., "Renormalization Group Analysis of the Small-World Network Model," *Physics Letters*, A, 263 (1999): 341; "Scaling and Percolation in the Small-World Network Model," *Physical Review*, E, 60 (1999): 7332. Thanks to the model's algorithmic simplicity, it is often preferred over the original Watts-Strogatz model by those trying to calculate the properties of the model.

THE FIFTH LINK: HUBS AND CONNECTORS

Page 55 Malcolm Gladwell, *The Tipping Point* (New York: Little, Brown, 2000).

Page 55 The phone book test used by Gladwell was invented by sociologists to estimate the number of social links people have. For a recent review on methods for measuring the size of a person's social links, see Linton C. Freeman and Claire R. Thompson, "Estimating Acquaintanceship Volume," in *The Small World*, ed. Manfred Kochen (Norwood, N.J.: Ablex, 1989), 147–158.

Page 57 On the technical level, we can easily find out how many outgoing links a given page has by simply visiting it and counting the URLs on it. It is harder, however, to count the number of incoming links. Incoming links represent the links from other Websites that point to a given document. For example, my graduate students have Webpages of their own, and they all have links to my Webpage. If you visit their page, with one click you can get to mine. However, when you visit my page, you have no way of knowing what other pages link to it. To find out how many incoming links my Webpage has, you have to visit each of the one billion pages that are on the Web and check if they have a link to my Website. The number of incoming links a Webpage has reflects its popularity: The more incoming links a page has, the more people must have visited and liked it. Most important, the more incoming links it has, the easier it is to find it by surfing the web. If nobody points at your Webpage, then for practical purposes it does not exist.

Page 57 Several search engines, such as Google or AltaVista, allow anybody to find pages that link to a given Webpage. You can reach this function by typing "link:"

followed by the URL into the search window. For example, for finding the links pointing to www.nd.edu/~networks you type link:http://www.nd.edu/~networks.

Page 57 For the study on the Notre Dame Website, see Réka Albert, Hawoong Jeong, and Albert-László Barabási, "Diameter of the World Wide Web," *Nature* 401 (1999): 130–131. The raw data giving information about which page is connected to which other pages, allowing you to reconstruct the network behind the Web sample and to determine how many hubs are out there, is available at http://www.nd.edu/~networks/database/index.html.

Page 58 For a summary of the 200-million-Webpage study see A. Broder, R. Kumar, F. Maghoul, P. Raghavan, S. Rajagopalan, R. Stata, A. Tomkins, J. Wiener, "Graph Structure in the Web," paper presented at the Ninth International World Wide Web Conference, http://www9.org/w9cdrom/160/160.html.

Page 58 The degree distribution, which allowed me to identify the connectivity of most of the connected nodes mentioned in the text when referring to the 200 million sample, was provided by A. Tomkins.

Page 58 The number of outgoing links, k_{out}, depends entirely on the Webpage's creator because he or she is the only one who can add them to the document. How many such links did we expect to see on a typical page? The power of the Web is that it is a hypertext, which allows the Webpage designer to structure the information into pages and subpages linked to the main page. Thus any decent book on Webpage design will warn us not to overcrowd our Webpage. The first page should be a roadmap, informative and easy to read. All details should go in the background, on extra pages, organized in a hierarchy of increasingly specific documents. How many links can we put on a page without making it too crowded? We typically have space for a few hundred words, including perhaps five to fifteen links. If everyone followed the advice of Web design books, most pages would have an optimal number of links, striking a healthy balance between maximum information content and readability. Of course, some pages could have more, others less than this optimal number, depending on the esthetic values of the Webmaster, but there would be a golden mean and very few gross violators. In this ideal world the distribution of links should follow a bell curve, or a Gaussian distribution in the mathematical language, peaked around an optimal k_{out}, very similar to the prediction of the random network model.

Page 58 The highly connected nodes on the World Wide Web are often called hubs and authorities, referring to nodes with many outgoing or incoming links, respectively. See, e.g., J. Kleinberg, "Authoritative Sources in a Hyperlinked Environment," *Proceedings of the 9th Association for Computing Machinery—Society for Industrial and Applied Mathematics. Symposium on Discrete Algorithms* (1998); extended version in *Journal of the* ACM, 46 (1999): 604-632.

Page 58 Craig Fass, Mike Ginelli, and Brian Turtle, *Six Degrees of Kevin Bacon* (New York: Plume, 1996).

Page 59 The Oracle of Bacon at Virginia can be found at http://www.cs.virginia.edu/oracle/.

Page 60 The important role of the actor hubs in Hollywood is illustrated by the fact that the shortest paths from Marilyn Monroe, Mike Meyer, or Charlie Chaplin to Bacon pass through the same actor, Robert Wagner. Wagner is the most important hub in Bacon's neighborhood, being his crucial link to historical Hollywood. Indeed,

Wagner has played in at least 101 movies and collected 2,017 links. Though he is not the most connected actor, he occupies the prominent twenty-fourth spot, a rank that would make Bacon rather envious.

Page 62 The ranking of Hollywood actors was based on a measurement Hawoong Jeong made in 2000 after downloading the IMDb.com database and reconstructing the network behind Hollywood. The results of similar measurements are often quoted in the press. Because the data were taken at different times, there could be minor differences regarding the precise order in which the most connected actors appear or the number of links they have. The measurements are consistent, however, regarding the identity of the most connected actors and their separation from the rest of Hollywood.

Page 63 For evidence of hubs in the molecular network of the cell, see, e.g., Chapter 13, and Hawoong Jeong, Bálint Tombor, Réka Albert, Zoltán N. Oltvai, and Albert-László Barabási, "The Large-Scale Organization of Metabolic Networks," *Nature* 407 (2000): 651; Hawoong Jeong, Sean Mason, Albert-László Barabási, and Zoltán N. Oltvai, "Centrality and Lethality of Protein Networks," *Nature* 411 (2001): 41–42; Andreas Wagner and David Fell, "The Small World Inside Large Metabolic Networks," *Proceedings of the Royal Society of London B*, Vol. 268 (Sept. 7, 2001): 1803-1810.

Page 63 For hubs in the Internet's topology see M. Faloutsos, P. Faloutsos, and C. Faloutsos, "On Power-Law Relationships of the Internet Topology," *Proceedings of ACM Special Interest Group on Data Communication (SIGCOMM), 1999* (Cambridge, Mass., Aug. 1999).

Page 63 For the phone call graph see J. Abello, P. M. Pardalos, and M. G. C. Resende, *Disc. Math. and Theor. Comp. Sci.*, DIMACS ser., 50 (1999): 119; William Aiello, Fan Chung, Linyuan Lu, *A Random Graph Model for Massive Graphs, Proceedings of the 32nd ACM Symposium on Theor. Comp.* (2000).

Page 63 Emanuel Rosen, *The Anatomy of Buzz* (New York: Doubleday, 2000).

Page 63 The FDR acquaintanceship network is discussed in H. Rosenthal, *Acquaintances and Contacts of Franklin Roosevelt* (master's thesis, Massachusetts Institute of Technology, 1960). See also Linton C. Freeman, and Claire R. Thompson, "Estimating Acquaintanceship Volume," 147–158, in *The Small World*, Edited by Manfred Kochen (Norwood: Ablex, NJ, 1989).

Page 63 For the discussion on hubs in the p53 network within the cell, see Bert Vogelstein, David Lane, and Arnold J. Levine, "Surfing the p53 Network," *Nature* 408 (2000): 307–310.

Page 64 For a discussion on keystone species, see Simon Levin, *Fragile Dominion* (Cambridge, Mass.: Perseus, 1999). For a discussion on hubs and keystone species, see Ricard V. Solé and José M. Montoya, *Complexity and Fragility in Ecological Networks*, http://www.santafe.edu/sfi/publications/Abstracts/00-11-060abs.html.

The Sixth Link: The 80/20 Rule

Page 65 The anecdote about Pareto is recalled in several places. See, e.g., the biographical note by Arthur Livingston in the English translation of *Trattato di Sociologia Generale, The Mind and Society* (New York: Harcourt Brace, 1942).

Page 66 The 80/20 law has generated numerous articles in the business world, eventually getting its own book. See Richard Koch, *The 80/20 Principle—The Secret to Success by Achieving More with Less* (New York: Currency, 1998).

Page 67 Our finding that a power law described the World Wide Web's topology was published in Réka Albert, Hawoong Jeong, and Albert-László Barabási, "Diameter of the World Wide Web," *Nature* 401 (1999): 130–131. An independent study arrived at the same conclusion. See R. Kumar, P. Raghavan, S. Rajalopagan, and A. Tomkins, "Extracting Large-Scale Knowledge Bases from the Web," *Proceedings of the 9th ACM Symposium on Principles of Database Systems* 1 (1999).

Page 68 In scientific articles the power-law nature of the degree distribution is often written in terms of probabilities. The probability that a randomly selected node has exactly k links follows $P(k) \sim k^{-\gamma}$, where γ is called the degree exponent.

Page 68 For an elementary introduction to power laws and their occurrence in various systems see Mark Buchanan, *Ubiquity: The Science of History . . . Or Why the World Is Simpler Than We Think* (New York: Crown Publishers, 2001). For power laws appearing in another much studied context, called self-organized criticality, see Per Bak, *How Nature Works* (Oxford, England: Oxford University Press, 1996).

Page 69 For the power-law nature of the Hollywood actor network see Albert-László Barabási and Réka Albert, "Emergence of Scaling in Random Networks," *Science* 286 (1999), 509–512; Albert-László Barabási, Réka Albert, and Hawoong Jeong, "Mean-Field Theory for Scale-Free Random Networks," *Physica* A, 272 (1999), 173–187.

Page 69 For power laws in science collaborations see A.-L. Barabási, H. Jeong, E. Ravasz, Z. Néda, T. Vicsek, A. Schubert, *On the Topology of the Scientific Collaboration Networks*, http://xxx.lanl.gov/abs/cond-mat/0104162. Similar results were obtained independently by Mark Newman, focusing on physicists, computer scientists, and other fields. See M. E. J. Newman, "The Structure of Scientific Collaboration Networks," *Proceedings of the National Academy of Sciences* 98 (2001): 404–409; "Scientific Collaboration Networks: I. Network Construction and Fundamental Results," *Physics Review*, E 64 (2001), 016131; "Scientific Collaboration Networks: II. Shortest Paths, Weighted Networks, and Centrality," *Physics Review*, E 64 (2001): 016132.

Page 69 For power laws within the cell see H. Jeong, B. Tombor, R. Albert, Z. N. Oltvai, and A.-L. Barabási, "The Large-Scale Organization of Metabolic Networks," *Nature* 407 (2000): 651–654; Hawoong Jeong, Sean Mason, Albert-László Barabási, and Zoltán N. Oltvai, "Lethality and Centrality in Protein Networks," *Nature* 411 (2001): 41–42; Adreas Wagner and David A. Fell, "The Small World Inside Large Metabolic Networks," *Proceeding of the Royal Society, London*, 268 (2001): 1803–1810.

Page 69 For the frequency of scientific citations see Sid Redner, "How Popular Is Your Paper? An Empirical Study of the Citation Distribution," *Euro. Phys. Journal*, B, 4 (1998), 131; see also S. Bilke and C. Peterson, "Topological Properties of Citation and Metabolic Networks," *Physical Review*, E 64 (2001): 036106.

Page 70 To be sure, the U.S. routing map is meticulously designed by the airlines to maximize profit. As a result, as Luis Amaral and colleagues from Boston University have shown, the distribution of the number of passengers visiting an airport has an exponential tail. Yet, since the network's topology is dominated by hubs, it has all the visual attributes of power law networks, therefore it offers an excellent example for remembering the main features of scale-free networks. For a study on the airline system, see L. A. N. Amaral, A. Scala, M. Barthélémy, and H. E. Stanley, "Classes of Small-World Networks," *Proceedings of the National Academy of Sciences* 97 (2000): 11149–11152.

Page 73 For an extensive but fascinating biography of water written for a general audience see Philip Ball, *Life's Matrix* (New York: Farrar, Straus and Giroux, 1999).

Page 74 For an excellent prerenormalization-group introduction into critical phenomena see the classic book by H. Eugene Stanley, *Introduction to Phase Transitions and Critical Phenomena* (Oxford, England: Oxford University Press, 1971).

Page 75 Note that, though both water and magnetic systems undergo a phase transition, they do so in very different ways. The water-ice phase transition is what physicists call a first-order phase transition, meaning that the relevant thermodynamic quantities change discontinuously at the transition point (i.e., they jump). Magnetic systems, in contrast, display a so-called second-order phase transition (i.e., the thermodynamic quantities change continuously at the transition point). It turns out that the theoretical tools required to describe these different phase transitions, though having the same roots, are rather different when we get to the heart of the matter. It is the second-order phase transition that always leads to power laws as we approach the critical point.

Page 75 Kadanoff's Christmas discovery in the context of the history of critical phenomena is recalled in Leo P. Kadanoff's, *From Order to Chaos, Essays: Critical, Chaotic and Otherwise* (Singapore: World Scientific, 1993): 157–163. Kadanoff does share the credit with others, as the scaling ideas that he introduced were independently discovered by several other researchers, including Michael Fisher, Ben Widom, A. Z. Patashinskii, V. L. Pokrovskii. It is believed that Wilson was awarded an unshared Nobel prize for critical phenomena precisely because there were too many credited with discovering the crucial prerequisite, the scaling concepts. Unfortunately the prize left several truly crucial contributions unhonored. Many of us feel that several other people's crucial contributions also deserve recognition.

Page 76 For Wilson's two seminal papers see Kenneth G. Wilson, "Renormalization Group and Critical Phenomena: I. Renormalization Group and the Kadanoff Scaling Picture," *Physics Review*, B, 4 (1971): 3174–3183; "Renormalization Group and Critical Phenomena. II. Phase-Space Cell Analysis of Critical Behavior," *Physical Review* B, 4 (1971): 3184–3205. For a recent pedagogic review of the field, see J. J. Binney, N. J. Dowrick, A. J. Fisher, and M. E. J. Newman, *The Theory of Critical Phenomena: An Introduction to the Renormalization Group* (Oxford, England: Oxford University Press, 1992).

Page 77 Second-order phase transitions are normally reversible, which means that we observe power laws independent of our direction as we cross the critical point, going from order to disorder, or the other way, from disorder to order.

Page 78 Universality became the guiding principle for understanding many disparate phenomena. It told us that down the line the laws of physics governing complex systems and the transition from disorder to order are simple, reproducible, and ubiquitous. We now know that the same universal mechanisms that generate the shapes of snowflakes also govern the shape of neurons in the retina. Power laws and universality emerge in economic systems, describing how companies grow and how cotton prices fluctuate. They explain how birds and fish flock and how earthquakes differ in magnitude. They are the guiding principle behind the two most intriguing discoveries of the second half of the twentieth century: chaos and fractals. As a result, a second revolution in statistical mechanics, which took place in the eighties and nineties, has focused on the important question of how power laws can appear in many different systems, some of which do not seem to undergo a phase transition. Self-organized criticality, a subfield of statistical mechanics, has united many researchers aiming to give a generic answer to this question.

Page 78 While it is relatively easy to give credit for scaling and renormalization group, it is much harder to catch the origin of the *universality* concept. Kadanoff used it a year after his seminal paper on scaling in a review article. He remembers hearing it during a conversation in a Moscow dollar bar with Sasha Polykov and Sasha Migdal, two prominent Russian physicists working at the border of phase transitions and field theory, a branch of physics commonly frequented by both particle physics and condensed matter physics. But it has been used within statistical mechanics by several others under different contexts. Little noticed, however, is the fact that universality has grown to take on a new meaning implying that the properties of rather different systems are identical while undergoing a transition from disorder to order.

THE SEVENTH LINK: RICH GET RICHER

Page 81 The paper written on the plane was published about five month later as Albert-László Barabási and Réka Albert, "Emergence of Scaling in Random Networks," *Science* 286 (1999): 509–512.

Page 82 Regarding the amount of information that will be stored on the Web in ten years, see Phil Bernstein, Michael Brodie, Stefano Ceri, David DeWitt, Mike Franklin, Hector Garcia-Molina, Jim Gray, Jerry Held, Joe Hellerstein, H. V. Jagadish, Michael Lesk, Dave Maier, Jeff Naughton, Hamid Pirahesh, Mike Stonebraker, and Jeff Ullman, "The Asilomar Report on Database Research," ACM *Sigmod Record* 27, no. 4 (1998): 74-80.

Page 82 The information regarding the growth of the Hollywood network is based on data collected by Hawoong Jeong from the IMDb.com database.

Page 83 For a detailed analysis of Model A see Albert-László Barabási, Réka Albert, and Hawoong Jeong, "Mean-Field Theory for Scale-Free Random Networks," *Physica* A, 272 (1999): 173–187.

Page 84 For online advertising budgets, see Michell Jeffers and Evanthei Schibsted, "The Sizzle: What's New and Now in Marketing and Advertising for E-Business and E-Commerce," *Business 2.0* (May 2000): 161–162.

Page 85 Of course, when it comes to news outlets, we all have similar preferences, navigating to the few sites that can afford to bring us smart, up-to-the-minute coverage. For less mainstream subjects, however, our choices are less predictable, and therefore, more random. Indeed, we might be the only people linking to our high-school friend's Webpage. But in the majority of cases we all follow an unconscious bias, linking with a higher probability to the more connected nodes.

Page 85 Direct quantitative evidence has been found for the existence of preferential attachment in networks as diverse as the Internet, Hollywood, collaboration networks, and citation networks. See H. Jeong, Z. Néda, A.-L. Barabási, "Measuring Preferential Attachment for Evolving Networks," http://xxx.lanl.gov/abs/cond-mat/0104131; M. E. J. Newman, "Clustering and Preferential Attachment in Growing

Networks," *Physical Review*, E 64, (2001): 025102; Ramualdo Pastor-Satorras, Alexei Vazquez, Alessandro Vespignani, "Dynamical and Correlation Properties of the Internet," *Physical Review Letters*, 87 (2001): 258701; K. A. Eriksen, and M. Hornquist, "Scale-Free Growing Networks Imply Preferential Attachment," *Physical Review*, E 65, (2001): 017102.

Page 86 The scale-free model, together with the two basic concepts, growth and preferential attachment, was introduced in Albert-László Barabási and Réka Albert, "Emergence of Scaling in Random Networks," *Science* 286 (1999): 509–512; Albert-László Barabási, Réka Albert, and Hawoong Jeong, "Mean-Field Theory for Scale-Free Random Networks," *Physica* A, 272 (1999): 173–187. Note that, though for simplicity we have chosen to link the new nodes to exactly two other nodes in the example discussed in this book, in general they can be linked to an arbitrary number of nodes without changing the model's basic features.

Page 88 The fact that the scale-free model generates a power law with a degree exponent equal to three has been proven exactly by some of Erdős's former collaborators. See B. Bollobás, O. Riordan, J. Spencer, G. Tusnády, "The Degree Sequence of a Scale-Free Random Graph Process," *Random Structures and Algorithms* 18 (May 2001): 279–290.

Page 89 The extension of the scale-free model to include internal links was published in Réka Albert and Albert-László Barabási, "Topology of Evolving Networks: Local Events and Universality," *Physical Review Letters* 85 (2000): 5234.

Page 89 The work of the Boston University group on aging was published in L. A. N. Amaral, A. Scala, M. Barthélémy, and H. E. Stanley, "Classes of Small-World Networks," *Proceedings of the National Academy of Sciences* 97 (2000): 11149–11152.

Page 89 The Porto group published two papers very close together. See S. N. Dorogovtsev, J. F. F. Mendes, "Evolution of Reference Networks with Aging," *Physical Review*, E 62 (2000): 1842; S.N. Dorogovtsev, F. F. Mendes, and A. N. Samukhim, "Structure of Growing Networks with Preferential Linking" *Physical Review Letters* 85 (2000): 4633.

Page 90 The effect of nonlinearities on the topology of scale-free networks, accounting for the fact that the attachment rate could be proportional to k^{γ}, was published in P. L. Krapivsky, S. Redner, and F. Leyvraz, "Connectivity of Growing Random Networks," *Physical Review Letters* 85 (2000): 4629–4632.

Page 90 For a detailed summary of the various extensions of the scale-free model and a general summary of the field of complex networks, see two recent review articles: S. N. Dorogovtsev and J. F. F. Mendes, "Evolution of Networks," *Advances in Physics* (in press, 2002); Réka Albert and Albert-László Barabási, "Statistical Mechanics of Complex Networks," *Reviews of Modern Physics* 74, (Jan. 2002): 47-97.

Page 92 The scale-free nature of language was addressed by several research groups. With Soon-Hyung Yook and Hawoong Jeong we connected all synonyms by links, finding that the network we obtained had a scale-free topology. We never published the results. Several groups, however, have published excellent papers analyzing the web within language, using different criteria for the links, each finding a scale-free topology. See, e.g., Ramon Ferrer i Cancho and Ricard V. Solé, "The Small-World of Human Language," *Proceedings of the Royal Society of London B*, 268 (2001): 2261–2265; Mariano Sigman and Guillermo Cecchi, Global Organization of the Wordnet Lexicon, Proceedings of the National Academy of Sciences, 99 (2002):

1742–1747. S.N. Dorogovtsev, J. F. F. Mendes, "Language as an Evolving Word Web," *Proceedings of the Royal Society of London B* 268 (Dec. 2001): 2603–2606.

The Eighth Link: Einstein's Legacy

Page 93 Yahoo! Inc. dropped Inktomi Corp. as its default search engine on June 26, 2000, and replaced it with Google. All major media outlets covered the event. The canceled deal was not expected to have an immediate impact on Inktomi's financial future, since at that time it had more than eighty customers and the Yahoo! partnership accounted for less than 2 percent of its revenue. Yet Inktomi's stock plunged over 25 5/16 points to close at 115 1/16 on NASDAQ.

Page 93 The Internet Archive's colloquium took place on March 8 and 9, 2000, at the Presidio, San Francisco, the home of the Archive. For more information on the Archive's goals see Chapter 12.

Page 94 For a detailed visual account of the birth of the Newton Apple handheld, see Doug Menuez, Markos Kounalakis, and Paus Saffo, *Defying Gravity: The Making of Newton* (Hillsboro, Ore.: Beyond Words, 1993). Unfortunately the book leaves off where things get interesting, as Newton hits the marketplace.

Page 94 For a detailed account of successful latecomers in business see Joan Indiana Rigdon, "The Second-Mover Advantage," *Red Herring*, September 1, 2000.

Page 96 The fitness model was published in G. Bianconi, A.-L. Barabási, "Competition and Multiscaling in Evolving Networks, *Europhysics Letters* 54 (May 2001): 436–442. For extensions of the model that include additional effects taking place in real networks such as additive or multiplicative fitness, see G. Ergun and G. J. Rodgers, *Growing Random Networks with Fitness*, *Physica A 303* (Jan. 2002): 261-272.

Page 98 For a detailed historical account of the Bose-Einstein relationship, as well as the birth of the ideas behind Bose-Einstein condensation, see William Blanpied, "Einstein as Guru? The Case of Bose," in *Einstein: The First Hundred Years*, ed. Maurice Goldsmith, Alan Mackay, and James Woudhuysen (Oxford, England: Pergamon Press, 1980), 93–99. See also Albrech Fölsing, *Albert Einstein: A Biography* (New York: Viking, 1997).

Page 100 A bizarre manifestation of Bose-Einstein condensation was discovered by Pyotr Kapitza and John F. Allen in 1938. Helium, the light gas used in blimps and birthday balloons, undergoes Bose-Einstein condensation below 2.2 Kelvin, becoming a superfluid. A striking manifestation of helium's new state is the loss of viscosity, allowing this liquid to slither up the walls and out of an open container. Imagine your morning coffee crawling up the mug's wall and slipping out. But that is exactly what helium does, a visible manifestation of Bose-Einstein condensation.

Page 100 For an elementary description of the new discoveries regarding Bose-Einstein condensation as well as their potential applications see, e.g., Graham P. Collins, "The Coolest Gas in the Universe," *Scientific American* (Dec. 2000): 92–99; Wolfgang Ketterle, "Experimental Studies of Bose-Einstein Condensation," *Physics Today* 52 (Dec. 1999): 30–35; Eric A. Cornell and Carl E. Wieman, "The Bose-Einstein Condensate," *Scientific American* (Mar. 1998): 40–45.

Page 101 The link between networks and Bose-Einstein condensation was published in G. Bianconi and A.-L. Barabási, "Bose-Einstein Condensation in Complex Networks," *Physical Review Letters*, 86 (June 2001): 5632–5635. Several researchers have extended this work, showing that the winner-takes-all phenomenon can be described without invoking the link to quantum mechanics. See, e.g., S. N. Dorogovtsev and J. F. F. Mendes, "Evolution of random networks," *Advances in Physics* (in press, 2002.)

Page 102 Bose-Einstein condensation is not the only area where the tools of quantum mechanics proved useful in the study of complex networks. The spectral properties of random matrices, an area pioneered by Eugene Wigner in the 1960s, was the starting point of studies on the spectral properties of complex networks. See for example I. J. Farkas, I. Derényi, A. L. Barabási, T. Vicsek, "Spectra of 'Real-World' Graphs: Beyond the Semi-Circle Law," *Physical Review* E 64 (2001): 026704; K. I. Goh, B. Kahng, D. Kim, "Spectra and Eigenvectors of Scale-Free Networks," *Physical Review* E 64 (2001): 051903. The tools of field theory, a mathematically advanced branch of quantum mechanics, has also found its applications in complex networks. See A. Krzgwicki, "Defining Statistical Ensembles of Random Graphs," http://www.laml.gov/abs/cond-mat/ 0110574; Z. Burda, J. D. Correia, A. Krzgwicki, "Statistical Ensemble of Scale-Free Random Graphs," *Physical Review* E 64 (2001): 046118.

Page 103 Note that a phenomenon very similar to Bose-Einstein condensation is predicted in networks with nonlinear preferential attachment. See P. L. Krapivsky, S. Redner, and F. Leyvraz, "Connectivity of Growing Random Networks," *Physical Review Letters*, 85 (2000): 4629–4632.

Page 104 The network describing the operation systems market is a so-called *bipartite graph*. Bipartite graphs are made of two different sets of nodes such that each node is allowed to link only to nodes belonging to the other set. Direct links between nodes belonging to the same set are forbidden. In the Microsoft example one set of nodes are the operating systems, and the other are the numerous consumers choosing (linking to) an operating system. A similar bipartite graph describes Hollywood as well. There one set of nodes are the actors, and the other are the movies in which they played. In this bipartite graph, actors are not linked to each other. Rather, each actor links to movies only. From this bipartite graph one can easily generate the actor network discussed in Chapter 5 by linking together all actors that point to the same movie. For a discussion of bipartite graphs see, e.g., M. E. J. Newman, S. H. Strogatz, and D. J. Watts, "Random Graphs with Arbitrary Degree Distributions and Their Applications," *Physical Review* E, 64, (2001): 026118; Steven H. Strogatz, "Exploring Complex Networks," *Nature* 410 (2001): 268–276.

Page 104 For a subjective history of operating systems see Neal Stephenson, *In the Beginning Was the Command Line*, http://www.cryptonomicon.com.

Page 105 The data regarding the market share of computer makers comes from IDC and is available at http://www.idc.com/.

Page 105 The operation systems market-share information comes from Stephanie Miles and Joe Wilcox, "Windows 95 Remains the Most Popular Operating System," *Cnet.com*, July 20, 1999.

Page 106 An example of a system where fitness had to be called in to explain the network topology is the Internet. See, e.g., Romualdo Pastor-Satorras, Alexei Vazquez, and Alessandro Vespignani, "Dynamical and Correlation Properties of the Internet," *Physical Review Letters*, 87 (2001): 258701.

THE NINTH LINK: ACHILLES' HEEL

Page 109 For a vivid description of the scene in Denver during the July 2, 1996, Denver electric power failure, see, e.g., L. M. Collins, "Power Grid Fails, Blackout Affects

1.5 Million in West," *Denver News-Times*, July 3, 1996; "Power Grid Fails, Blackout Affects Millions in West," Nando.net, July 2, 1996. For the August repeat of the same event, see *Sagging power lines, hot weather blamed for blackout*, CNN, August 11, 1996.

Page 110 The various electric power breakdowns hitting the United States, as well as the vulnerability of the electricity network were widely discussed both in the popular press and by various professional organizations. See, e.g., Massoud Amin, "Toward Self-Healing Infrastructure Systems," *IEEE Computer* (Aug. 2000): 2–11; D. N. Kosterev, C. W. Taylor, and W. A. Mittlestadt, "Model Validation of the August 10, 1996 WSCC System Outage," *IEEE Transactions on Power Systems* 14 (Aug. 1999): 967–977. For a general discussion on deregulation and its impact on the infrastructure, see Alan Weisman, "Power Trip: The Coming Darkness of Electricity Deregulation," *Harper's*, Oct. 2000, 76–85.

Page 111 For a discussion of the number of species on Earth, biodiversity, and extinctions, see, e.g., Robert M. May, "How Many Species Inhabit the Earth?" *Scientific American*, Oct. 1992, 42–48; Joel L. Swerdlow, "Biodiversity: Taking Stock of Life," *National Geographic* 192 (Feb. 1999): 2–41; Virginia Morell, "The Sixth Extinction," *National Geographic* 192 (Feb. 1999): 43–59.

Page 111 A much-cited source of nature's robustness is *redundancy*, a property inherent in all networks but virtually absent from most human designs. In most networks there are a huge number of alternate paths between most pairs of nodes. Indeed, though the local senator might offer a short path to the president of the United States, he or she is not indispensable. Should he or she decline to introduce us, there are many other paths, often equally short, that link us to the president. A similar redundancy is built into the Internet. If a router does not work, messages will be rerouted along alternative paths. Redundancy is present in ecosystems as well. A predator only very rarely feeds only on one species. Indeed, upon the successful elevation of mice to pet status in the household, the cat grudgingly survives on canned food. Alternative routes are an important source of redundancy and error tolerance. That is, in most natural systems the malfunction of several nodes is not fatal, because the paths eliminated by their absence can be replaced by one of the many alternative routes. Many of us have experienced this phenomenon when choosing an alternative route to our destination after the radio announces congestion along the shortest route or when bad weather or a cancelled flight reroutes us to a different hub during air travel. But could there be something beyond redundancy when it comes to robustness?

Page 112 The breakdown of a random network under random node removal is an inverse percolation problem. Percolation theory tells us that the transition from a fragmented to a fully connected network is a second-order phase transition. For a review of percolation see D. Stauffer and A. Aharony, *Introduction to Percolation Theory* (London: Taylor and Francis, 1994); A. Bunde and S. Havlin, eds., *Fractals and Disordered Systems* (Berlin: Springer, 1996); idem, eds., *Fractals in Science*, (Berlin: Springer, 1995).

Page 113 Our error tolerance study showing that scale-free networks are not vulnerable to attacks was published in Réka Albert, Hawoong Jeong, and Albert-László Barabási, "Attack and Error Tolerance of Complex Networks," *Nature* 406 (2000): 378. For a summary and a perspective, see the accompanying News & Views article, Yuhai Tu, "How Robust Is the Internet?" *Nature* 406 (2000): 353–354.

Page 113 For a discussion of the Internet's stability see, e.g., Craig Labovitz, Abha Ahuja, and Farnam Jahanian, "Experimental Study of Internet Stability and Wide-Area Backbone Failures," *Proceedings of Institute of Electrical and Electronics Engineers (IEEE) Symposium on Fault-Tolerant Computing FTCS* (Madison, Wis.: June 1999).

Page 114 In considering robustness we cannot ignore the dynamic properties of complex systems. Most systems known to be robust have numerous controls and feedback loops to ensure that they survive errors and failures. Indeed, Internet protocols were carefully designed to "route around the trouble," avoiding routers that malfunction; cells have numerous feedback mechanisms to correct errors, dismantling faulty proteins and shutting down malfunctioning genes. Our computer simulations indicated a new component to error tolerance, however. We learned that nature has carefully selected the structure of most complex systems, offering them an unparalleled degree of error and failure tolerance. By virtue of their topology only, these systems display a high degree of resilience, a property we called *topological robustness*.

Page 115 The calculation of the percolation threshold of scale-free networks was published in Reuven Cohen, Keren Erez, Daniel ben-Avraham, and Shlomo Havlin, "Resilience of the Internet to Random Breakdowns," *Physical Review Letters* 85 (2000): 4626. Similar results were obtained independently in D. S. Callaway, M. E. J. Newman, S. H. Strogatz, and D. J. Watts, "Network robustness and fragility: Percolation on random graphs," *Physical Review Letters* 85 (2000): 5468–5471.

Page 115 For a detailed list of links on MafiaBoy see http://www.mafiaboy.com.

Page 115 Note that not everybody believes that Operation Eligible Receiver did really exist. Some critics maintain that it is nothing more than a Pentagon ghost story, repeated often to journalists. See, e.g., http://www.soci.niu.edu/~crypt/other/eligib.htm.

Page 116 According to the Computer Currents Internet Dictionary (http://www.computeruser.com/resources/dictionary/dictionary.html) a *cracker* is "a person who breaks into computer systems, using them without authorization, either maliciously or just to show off." In contrast, a *hacker* is "one who is knowledgeable about computers and creative in computer programming, usually implying the ability to program in assembly language or low-level languages. A hacker can mean an expert programmer who finds special tricks for getting around obstacles and stretching the limits of a system." Therefore, to distinguish from the well-meaning hackers, we use the term *cracker* for those whose aim it is to launch a malicious attack against the Internet. For a more detailed discussion see, e.g., Pekka Himanen, *The Hacker Ethic and the Spirit of the Information Age* (New York: Random House, 2001); Richard Power, *Tangled Web: Tales of Digital Crime from the Shadows of Cyberspace* (Indianapolis, Ind.: Que, 2000); Steven Levy, *Hackers: Heroes of the Computer Revolution* (New York: Penguin Books, 1994).

Page 116 The fragility of networks against attacks was first discussed in the same publication as the error tolerance, i.e., Réka Albert, Hawoong Jeong, and Albert-László Barabási, "Attack and Error Tolerance of Complex Networks," *Nature* 406 (2000): 378.

Page 117 For an analytical approach to the attack problem see, e.g., Reuven Cohen, Keren Erez, Daniel ben-Avraham, and Shlomo Havlin, "Breakdown of the Internet under Intentional Attack," *Physical Review Letters* 86 (2001): 3682; and D. S. Callaway, M. E. J. Newman, S. H. Strogatz, and D. J. Watts, "Network robustness and fragility: Percolation on random graphs," *Physical Review Letters* 85 (2000): 5468–5471.

Page 117 For the resilience of the protein network to mutations and drug attacks, see Hawoong Jeong, Sean Mason, Albert-László Barabási, and Zoltán N. Oltvai, "Lethality and Centrality in Protein Networks," *Nature* 411 (2001): 41–42.

Page 117 For a discussion of the breakdown of the ecosystems under the removal of keystone species, see, e.g., Ricard V. Solé and José M. Montoya, *Complexity and Fragility in Ecological Networks*, http://xxx.lanl.gov/abs/cond-mat/0011196; and Ferenc Jordán and István Scheuring, "Can Keystones Help in Background Extinction?" (preprint, 2000). For the effect of human activities on the stability and potential breakdown of the ecosystem, see, e.g., Stuart L. Pimm and Peter Raven, "Biodiversity: Extinction by Numbers," *Nature* 403 (2000): 843–845. For protecting our biodiversity, see Norman Myers, Russell A. Mittermeier, Cristina G. Mittermeier, Gustavo A. B. da Fonseca, and Jennifer Kent, "Biodiversity Hotspots for Conservation Priorities," *Nature* 403 (2000): 853–858. For a phenomenal photographic journey into these hotspots, see Russell A. Mittermeier, Norman Myers, Patricio Robles Gil, and Cristina G. Mittermeier, *Hotspots: Earth's Biologically Richest and Most Endangered Terrestrial Ecoregions* (Mexico City: Cemex Conservation International, 2000).

Page 118 The phrase *Achilles' heel*, as applied to networks and the Internet, was suggested to me by Janet Kelley after I explained to her our results. Originally included in the title of our *Nature* paper, the phrase appeared only on the magazine's cover.

Page 118 The concept of keystone species was introduced by Robert Paine, in R. T. Paine, "A Note on Trophic Complexity and Community Stability," *American Naturalist*, 103 (Jan.–Feb. 1969): 91–93. For a general discussion on the sea otter, see chapter 1 in Simon Levin, *Fragile Dominion: Complexity and the Commons* (Cambridge, Mass.: Perseus, 1999).

Page 119 For a detailed discussion of the 1996 summer electricity breakdown, see D. N. Kosterev, C. W. Taylor, and W. A. Mittlestadt, "Model Validation of the August 10, 1996, WSCC System Outage," *IEEE Transactions on Power Systems* 14 (Aug. 1999): 967–977.

Page 121 For a discussion of cascading failures, see, e.g., Duncan J. Watts, *A Simple Model of Fads and Cascading Failures*, http://www.santafe.edu/sfi/publications/Abstracts/00-12-062abs.html.

The Tenth Link: Viruses and Fads

Page 123 The story of Gaetan Dugas as Patient Zero is recalled in Randy Shilts, *And the Band Played On* (New York: St. Martin's Press, 2000), a moving and frightening day-to-day chronicle of the AIDS epidemic. For an up-to-date status report on the AIDS epidemic see "Nature Insight—AIDS," *Nature* 410, no. 9 (2001): 961–1007.

Page 124 The story of Mike Collins and the Florida ballots cartoon is told by Collins to Robb Mandelbaum in "Only in America," *New York Times Magazine*, Nov. 26, 2000.

Page 126 For information on epidemics and diseases see Rob DeSalle, ed., *Epidemic! The World of Infectious Disease* (New York: New Press, 1999).

Page 126 Note that there are important differences between the various diffusion processes, such as the spread of ideas or viruses. For example, with many diseases you could be cured, unable to transmit the virus further, or could develop an immunity so that you were unable to be infected again. With ideas, though you may reject them, once you accept them you will continue to propagate them. Also, some of the viruses, such as

Ebola, would quickly kill their hosts, offering only a short time frame to pass them on, a phenomenon that is again absent from the spread of most fads and ideas. Despite these differences, many of the fundamental features of the spreading of fads and biological and computer viruses are rather similar, and thus we will often lump them together.

Page 127 The Iowa farmers study is published in Bryce Ryan and Neal C. Gross, "The Diffusion of Hybrid Seed Corn In Two Iowa Communities," *Rural Sociology* 8, no. 1 (1943): 15–24.

Page 128 For a simple description of the bell curve and its impact on buzz and marketing, see Emanuel Rosen, *The Anatomy of Buzz* (New York: Doubleday, 2000), 94–95.

Page 128 For the physician study see James Coleman, Elihu Katz, and Herbert Menzel, "The Diffusion of an Innovation Among Physicians," *Sociometry* 20, no. 4 (1957): 253–270). For an earlier study on opinion leaders, see Elihu Katz and Paul F. Lazarsfeld, *Personal Influence: The Part Played by People in the Flow of Mass Communications* (Glencoe, Ill., Free Press, 1955).

Page 131 For threshold models see Mark Granovetter, "Threshold Models of Collective Behavior," *American Journal of Sociology* 83, no. 6 (1978): 1420–1443. For a general review of the subject, see also Thomas W. Valente, *Network Models of the Diffusion of Innovations* (Cresskill, N.Y.: Hampton Press, 1995); Eric Abrahamson and Lori Rosenkopf, "Social Network Effects on the Extent of Innovation Diffusion: A Computer Simulation," *Organization Science* 8, no. 3 (1997): 289–309.

Page 132 For a general audience diary of the Love Bug see Lev Grossman, "Attack of the Love Bug," *Time*, May 15, 2000.

Page 133 Computer Viruses carried by email, such as Love Bugs, spread on a social network whose nodes are email users, connected if they used to send emails to each other. Recently, German scientists have shown that this network is scale-free. See Holger Ebel, Latz-Ingo Mielsch, Stefan Bornholdt, Scale-Free Topology of Email Networks, http://xxx.lanl.gov/abs/cond-mat/0201476.

Page 134 For a general article on computer virus spreading see Jeffrey O. Kephart, Gregory B. Sorkin, David M. Chess, and Steve R. White, "Fighting Computer Viruses," *Scientific American* (Nov. 1997): 88–93. For more detailed approaches, see Jeffrey O. Kephart, Gregory B. Sorkin, William C. Arnold, David M. Chess, Gerald J. Tesauro, and Steve R. White, "Biologically Inspired Defenses Against Computer Viruses," in *Machine Learning and Data Mining: Methods and Applications*, ed. R. S. Michalski (New York: John Wiley, 1998); Steve R. White, "Open Problems in Virus Research," *International Virus Bulletin* (Munich, Germany, Oct. 22–23, 1998).

Page 135 The absence of a threshold in spreading on a scale-free network was described in Romualdo Pastor-Satorras and Alessandro Vespignani, "Epidemic Spreading in Scale-Free Networks," *Physical Review Letters* 86 (2001): 3200–3203; *Epidemic Dynamics and Endemic States in Complex Networks, Physical Review*, E 63 (2001): 066117. For a perspective on these results see Alun L. Lloyd and Robert M. May, "How Viruses Spread Among Computers and People," *Science* 292 (2001): 1316.

Page 137 Fredrik Liljeros, Christofer R. Edling, Luis A. Nunes Amaral, H. Eugene Stanley, and Yvonne Aberg, "The Web of Human Sexual Contacts," *Nature*, 411 (2001): 907–908. The story behind the Stockholm-Boston discovery was revealed to me by Fredrick Liljeros and Christopher R. Edling, the first two authors of the study.

Page 137 The Trieste study predicts that the threshold for diffusion in a scale-free network vanishes when the degree exponent, γ, is smaller than three. For exponents larger than this critical value that threshold reemerges, and the behavior is similar to that seen in the random networks, i.e., less contagious viruses will die out. The exponents obtained by the Stockholm-Boston collaboration do not give unique guidance in this respect. For the one-year data they obtained $\gamma = 3.54 \pm 0.2$ (females) and $\gamma = 3.31 \pm 0.2$ (males), while the more extensive (but potentially more biased as well) data on all partners gives $\gamma = 3.1 \pm 0.3$ (females) and $\gamma = 2.6 \pm 0.3$ (males). Whereas the first two exponents are clearly larger than three, the latter two are smaller or are at the border. Clearly, more extensive surveys are needed to obtain a definitive answer.

Page 138 The famous 20,000 line comes from Wilt Chamberlain, *A View from Above* (New York: Villard Books, 1991).

Page 138 For a survey of the two-decades-long AIDS epidemic and its impact on society see, e.g., Sharon Begley, "AIDS at 20," *Newsweek*, June 11, 2001.

Page 138 For a detailed description of the laws governing the spreading of viruses, see Martin A. Nowak and Robert M. May, *Virus Dynamics: Mathematical Principles of Immunology and Virology* (Oxford, England: Oxford University Press, 2000).

Page 140 Strictly speaking, our simulations assumed that the hubs, once treated, could be again infected if they came into contact with an another infected node. As the situation stands today, those treated with the currently available drugs will indeed have a smaller virus count and are thus less likely to spread the disease. When it comes to AIDS, there are many other details that one needs to keep in mind. Our goal was only to show that the epidemic threshold will return if one focuses on the hubs. Therefore, irrespective of the details of the spreading process, curing the hubs would be a far more effective policy than random drug distribution, assuming that the number of cures is limited. The goal would be, of course, to get the cure to everyone who needs it. For our study, see Zoltán Dezső, Albert-László Barabási, *Can We Stop the AIDS Epidemic?* http://xxx.lane.gov/abs/cond-mat/ 0107420. See also Romualdo Pastor-Satorras, Allesandro Vespignani, "Immunization of Complex Networks," *Physical Review*, E 65 (2002): 036104.

THE ELEVENTH LINK: THE AWAKENING INTERNET

Page 143 The Baran quote comes from John Naughton, *A Brief History of the Future* (Woodstock, NY: Overlook Press, 2000), 93. For more on Baran's life see Chapter 6 in Naughton's book.

Page 144 Recently a number of books and articles have focused on the history of the Internet. In addition to the above-cited Naughton book, see, e.g., James Gillies and Robert Cailliau, *How the Web Was Born: The Story of the World Wide Web* (Oxford, England: Oxford University Press, 2000). The latter focuses mainly on the Web but covers some elements of Internet history as well.

Page 145 Paul Baran's historical RAND memoranda are available on the web at http://www.rand.org/publications/RM/baran.list.html. The one interesting to us regarding the network's topology is Paul Baran, *Introduction to Distributed Communications Networks*, RM–3420-PR, available on the same Web page. Figure 11.1 comes from this paper.

Page 146 Note that the genesis of packet switching has been disputed lately. Some believe that Leonard Kleinrock also arrived at the idea independently of Baran and

Davies. For a discussion of the historical dispute see Katie Hafner, "A Paternity Dispute Divides Net Pioneers," *New York Times*, Nov. 8, 2001.

Page 147 As James Gillies and Robert Cailliau discuss in the above cited *How the Web Was Born*, despite all appearances, the working principle of the Internet is closer to that of the post office than the telephone network. In the traditional analog telephone system, when you call somebody your phone is connected through a succession of wires and switches directly to the phone of the person with whom you wish to speak. Once the connection is established, a physical line is apportioned entirely to serving the two of you, inaccessible to anyone else's call whether you say anything or not. The postal system works on a different principle. Post offices are connected by a network of roads. Letters collected by each office are sorted according to their destination and placed on trucks and planes that carry all letters that share the same route. In contrast to the phone system, you would never have a single truck take your letter directly from your house to its destination. Similarly, on the Internet, computers communicate by breaking messages into tiny parcels called packets. Just like a letter, each packet contains information about its destination. Each time a router receives a packet, it looks at the address and sends it to the router closest to its destination. Packets of the same message often travel different routes, since there are several alternate paths connecting any source and destination. When all the packets arrive, the recipient computer reassembles them, creating the e-mail message or the Webpage displayed on your screen.

Page 148 For a better idea of the work of the CAIDA collaboration see, e.g., K. C. Claffy, T. Monk, and D. McRobb, "Internet tomography," *Nature* (Jan. 1999) available at http://www.nature.com/nature/webmatters/.

Page 148 For a detailed and colorful description of the efforts behind mapping the Web and the Internet, see Martin Dodge and Rob Kitchin, *Atlas of Cyberspace* (New York: Addison-Wesley, 2002). See also Martin Dodge and Rob Kitchin, *Mapping Cyberspace* (London: Routledge, 2000).

Page 148 For a general discussion on the effect of self-organization on the Internet's topology see, e.g., Albert-László Barabási, "The Physics of the Web," *Physics World* (July 2001): 33-38.

Page 149 For the birth of email see John Naughton, *A Brief History of the Future* (Woodstock, NY: Overlook, 2000).

Page 150 Vern Paxson and Sally Floyd, "Why We Don't Know How To Simulate the Internet," *Proceedings of the 1997 Winter Simulation Conference*, ed. S. Andradottir, K. J. Healy, D. H. Withers, and B. L. Nelson. The "Sucess disaster" phrase also comes from this paper (see pg. 149).

Page 150 The discovery that the Internet has a power-law degree distribution was reported in M. Faloutsos, P. Faloutsos, and C. Faloutsos, "On Power-Law Relationships of the Internet Topology," [ACM SIGCOMM 99, comp.] *Computer Communications Review* 29 (1999): 251. For more recent measurements, confirming this finding on much larger samples, see, e.g., R. Succ and H. Tangmunarunkit, "Heuristics for Internet Map Discovery," *Proceedings of Infocom* (March 2000).

Page 151 For the timeline of the first Internet nodes, see John Naughton, *A Brief History of the Future*.

Page 152 Note that the first Internet model was introduced in Bernard M. Waxman, "Routing of Multipoint Connections," *IEEE Journal on Selected Areas in Communications*

6 (Dec. 1988): 1617-1622. Waxman laid down a large number of nodes on a plane and connected them randomly to each other. So far, this was no different from the random model of Erdős and Rényi. However, aware of the high costs of wiring, he wanted to discourage very long links. Therefore, he assumed that the probability that two nodes on the Internet are linked decrease exponentially with the distance between them. Waxman's simple model has dominated Internet modeling for decades. It was questioned only in 1999, when the Internet's scale-free nature was uncovered. But the exponential distance dependence made its way, together with growth and preferential attachment, into more modern Internet models as well. First, simulations indicate that with such a drastic distance dependence as the Waxman model offers, scale-free networks cannot develop. Second and even more important, Yook and Jeong's measurements indicate that the probability of connecting two nodes at distance *d* from each other decreased linearly with *d*—much weaker dependence than the exponential form assumed by Waxman.

Page 152 The presence of preferential attachment on the Internet is discussed in several publications. See, e.g., Soon-Hyung Yook, Hawoong Jeong, and Albert-László Barabási, *Modeling the Internet's Large-Scale Topology*, http://xxx.lanl.gov/abs/cond-mat/0107417; Hawoong Jeong, Zoltán Néda, Albert-László Barabási, *Measuring Preferential Attachment for Evolving Networks*, http://xxx.lanl.gov/abs/cond-mat/0104131; Romualdo Pastor-Satorras, Alexei Vazquez, Alessandro Vespignani, "Dynamical and Correlation Properties of the Internet," *Physical Review Letters*, 87 (2001): 258701.

Page 152 Assuming that nodes pop up proportionally to the population density and that the probability that a node will link to an another node with k links located at a distance r is proportional to k/r^{σ}, where σ is a free parameter that allows us to tune the effect of the spatial component: If σ is large, then distance is very important, whereas if $\sigma = 0$, then only preferential attachment matters for the Internet's evolution. The simulations offered a rather clear answer: As long as σ is smaller than two, a scale-free network emerges. However, if $\sigma > 2$, then the restricting effect of distance wins, and the network develops an exponential degree distribution. Our measurements have clearly indicated that for the Internet $\sigma = 1$, explaining why, despite the expenses of laying down the longer cables to get more bandwidth, the scale-free topology survives. Beyond explaining why the Internet is a scale-free network, these results also indicated how important it is to uncover, in quantitative terms, the different competing principles that govern network evolution. See Soon-Hyung Yook, Hawoong Jeong, and Albert-László Barabási, *Modeling the Internet's Large-Scale Topology*, http://xxx.lanl.gov/ abs/cond-mat/0107417.

Page 152 For other recent Internet models that incorporate the Internet's scale-free topology, see Alberto Medina, Ibrahim Matta, and John Byers, "On the Origin of Power Laws in Internet Topologies," [ACM SIGCOMM] *Computer Communications Review* 30, no. 2 (2000): 18–28); G. Caldarelli, R. Marchetti, and L. Pietronero, "The Fractal Properties of the Internet," *Europhysics Letters* 52 (2000): 386; K.I. Goh, B. Kahng, D. Kim, *Universal Behavior of Load Distribution in Scale-Free Networks, Physical Review of Letters*, 87 (2001): 278701; A. Capocci, G. Caldarelli, R. Marchetti, L. Pietronero, "Growing Dynamics of Internet Providers," *Physical Review*, E 64 (2001): 035101.

Page 152 Fractals, self-similar objects with nontrivial geometrical properties, were introduced by Benoit Mandelbrot. Subsequently, they were found to describe many natural objects, from snowflakes to cell colonies. See, e.g., B. Mandelbrot, *The Fractal Geometry of Nature* (New York: W. H. Freeman, 1977). For a more recent review, see T. Vicsek, *Fractal Growth Phenomena* (Singapore: World Scientific, 1992).

Page 153 The routing failure at MAI was described in several news accounts. See, e.g., "Router Glitch Cuts Net Access," *CNET*, April 25, 1997.

Page 155 For a discussion on the Code Red worm, see Carolyn Meinel, "Code Red for the Web," *Scientific American*, October 2001, 42–51.

Page 155 For parasitic computing, see Albert-László Barabási, Vincent W. Freeh, Hawoong Jeong, and Jay B. Brockman, "Parasitic Computing," *Nature* 412 (2001): 894–897. For further information, see http://www.nd.edu/~parasite/.

Page 157 For a detailed discussion of distributed computing, see, e.g., Ian Foster, "Internet Computing and the Emerging Grid," *Nature* (Dec. 2000), available at http://www.nature.com/nature/webmatters.

Page 158 For a truly interesting discussion on the electronic skin developing around the Earth see Neil Gross, "The Earth Will Don and Electronic Skin," Business Week (August 30, 1999): 68-70.

The Twelfth Link: The Fragmented Web

Page 162 For the reaction of the search engines on the NEC study see Thomas E. Weber's, "Fast Forward: Media in Motion," *Wall Street Journal*, April 3, 1998.

Page 163 For Inquirus, see Steve Lawrence and C. Lee Giles, "Inquirius: The NECI Meta Search Engine," Seventh International World Wide Web Conference, Brisbane, Australia (Amsterdam: Elsevier Science, 1998), 95–105.

Page 163 The findings of the NEC group have been published in two papers: Steve Lawrence and C. Lee Giles, "Searching the World Wide Web," *Science* 280 (1998): 98–100; and Steve Lawrence and C. Lee Giles, "Accessibility of Information on the Web," *Nature* 400 (1999): 107–109.

Page 164 For a detailed discussion of the size of the Web, see http://searchengine.com. For the latest statistics on the search engine sizes see Danny Sullivan's "Search Engine Sizes" *Search Engine Report*, August 15, 2001, available at http://searchengine.com/reports/sizes.html; see also "Numbers, Numbers—But What Do They Mean?" *Search Engine Report*, March 3, 2000, http://searchengine.com/sereport/00/03-numbers.html.

Page 166 The fragmented structure of the Web is known as the bow-tie theory and was first observed in A. Broder, R. Kumar, F. Maghoul, P. Raghavan, S. Rajagopalan, R. Stata, A. Tomkins, and J. Wiener, "Graph structure in the Web," Ninth International World Wide Web Conference, Amsterdam, http://www9.org/w9cdrom/160/160.html.

Page 168 As they age and become known, Webpages naturally travel across the continents. Their position is jointly determined by the creator of the Webpage and by the interest of the World Wide Web community in the pages' content. As links and Webpages are constantly added, removed, modified, and enriched and robbed of links, the population of these continents are in constant flux, compared to which the big influxes from Europe to the United States at the end of the twentieth and the beginning of this century are small, negligible events. A single well-placed link can determine the fate and location of thousands of Webpages, the whole landscape being reorganized by small or huge avalanches.

Page 169 For recent research on the properties of directed networks see S. N. Dorogovtsev, J. F. F. Mendes, and A. N. Samukhin, *Giant Strongly Connected Component of Directed Networks*, http://xxx.lanl.gov/abs/cond-mat/0103629; M. E. J. Newman, S. H. Strogatz, and D. J. Watts, "Random Graphs with Arbitrary Degree Distributions and Their Applications," *Physical Review*, E 64 (2001): 026118; B. Tadic, "Dynamics of Directed Graphs: the World Wide Web," *Physica*, A 293 (2001): 273–284.

Page 170 Cass R. Sunstein, *Republic.com* (Princeton, Princeton University Press, 2001).

Page 171 Justice Stewart's quote on pornography is cited in many places. See, e.g., "The Task of Defining What's Too Explicit to Be Seen," *USA Today*, Jan. 26, 1999, available at http://www.usatoday.com/life/cyber/tech/ctb114.htm.

Page 171 For the NEC work on communities on the Web, see Gary William Flake, Steve Lawrence, and C. Lee Giles, "Efficient Identification of Web Communities," *Proceedings of the Sixth International Conference on Knowledge Discovery and Data Mining* (Boston, Mass.: ACM Special Interest Group on Knowledge Discovery in Data and Data Mining, August 2000), 156–160. Several other groups have worked on very similar problems. See David Gibson, Jon Kleinberg, and Pranhakar Raghavan, "Inferring Web Communities from Link Topology," *Proceedings of the 9th ACM Conference on Hypertext and Hypermedia* (1998); and Ray R. Larson, *Bibliometrics of the World Wide Web: An Exploratory Analysis of the Intellectual Structure of Cyberspace*, http://sherlock.berkeley.edu/asis96/asis96.html.

Page 171 For a discussion of NP complete problems, see M. Garey and D. S. Johnson, *Computers and Intractability: A Guide to the Theory of NP-Completeness* (San Francisco: H. W. Freeman, 1979).

Page 172 Lada A. Adamic, "The Small World Web," *Proceedings of ECDL'99*, LNCS 1696 (Springer, 1999), 443–452. See also Lada A. Adamic and Eytan Adar, *Friends and Neighbors on the Web*, http://www.hpl.hp.com/shl/papers/web10/.

Page 173 Lawrence Lessig, *Code and Other Laws of Cyberspace* (New York: Basic Books, 1999).

Page 176 To know more about the Internet Archives, visit their Website at http://www.archive.org/.

Page 177 Most of our creative life is turning towards the Web. The modern photographer uses a digital camera and manipulates the bits to better express his vision. Some of these pictures will be printed and displayed in galleries. Most, however, are available only in electronic format on the Web. The bulk of poems are not published in anthologies any longer; they are available in Web archives. The Web is the primary medium for an increasing number of visual artists whose work cannot be appreciated without a browser. Yet all of this will be irreversibly lost thanks to badly curated sites, broken computers, and vanishing resources. We will not have Van Goghs in the future because their work, if unappreciated by their contemporaries, will not survive for future generations. The creative geniuses of the online world will be tossed out with computer upgrades or technology changes, not in centuries but in a few short years.

There is only one way to halt this tragic loss of history and creativity housed in the Web these days. We must archive everything out there for the generations to come. I believe that we should make a serious, perhaps government-supported effort to go well beyond the goals and possibilities of the Internet Archives and map the *full* Web. Every single page of it. We should make its past and current content instantly available to anybody anywhere.

Page 178 While in this chapter we focused mainly on the topology of the Web, recently a series of results have investigated our surfing patterns and our dynamical behavior on the Web, finding further evidence of emerging behavior and power laws. See Bernardo A. Huberman, Peter L. T. Pirolli, James E. Pitkow, and Rajan M. Lukose, "Strong Regularities in World Wide Web Surfing," *Science* 280 (1998): 95–97; Anders Johansen and Didier Sornette, *Download Relaxation Dynamics on the WWW Following Newspaper Publication of URL*, http://xxx.lanl.gov/abs/cond-mat/9907371; and Bernardo A. Huberman, *The Laws of the Web* (Cambridge, Mass.: MIT Press, 2001).

The Thirteenth Link: The Map of Life

Page 179 For leading causes of death in the United States (including depression), see the Website of the Centers for Disease Control and Prevention (CDC), http://webapp.cdc.gov/.

Page 179 For research on manic depression specifically, see Nick Craddock and Ian Jones, "Molecular Genetics of Bipolar Disorder," *British Journal of Psychiatry* 174, suppl. 41 (2001): 128–133. For research on depression with discussion on manic-depression as well, see Charles B. Nemeroff, "The Neurobiology of Depression," *Scientific American*, June 1998, 42.

Page 180 The decoding of the human genome has been covered widely by the press, both for the occasion of the official White House announcement on June 25, 2000, as well as the publication of the genome on February 15 and 16, 2001. See *Science* 291 (Feb. 2001), and *Nature* 409 (Feb. 2001).

Page 181 For a recent discussion of postgenomic biology and changing views of the role of the gene, see Evelyn Fox Keller, *The Century of the Gene* (Cambridge, Mass.: Harvard University Press, 2000).

Page 181 For more insight into the increasing role of networks in understanding the cell, see J. Craig Venter et al, "The Sequence of the Human Genome," *Science* 291 (2001): 1304–1351, especially 1347–1348.

Page 182 For an excellent introduction to cell biology see Bruce Alberts, Dennis Bray, Julian Lewis, Martin Raff, Keith Roberts, and James D. Watson, *Molecular Biology of the Cell* (New York: Garland, 1994).

Page 182 The research on metabolism goes back to the nineteenth century, responding to the French wine maker's need to control the steps by which yeast cells change glucose into alcohol and bubbles of carbon dioxide. This ancestry is preserved in the name *enzyme*, whose root means "in yeast." Therefore, biochemistry can be viewed as a giant mapping project struggling to create an inventory of all possible chemicals and reactions present in the cell. For a detailed history of mapping out metabolism, see Horace Freeland Judson, *The Eighth Day of Creation: Makers of the Revolution in Biology* (Plainview, NY: Cold Spring Harbor Laboratory Press, 1996).

Page 182 Note that various subnetworks, such as the metabolic or the protein interaction network, are not independent of one another. Indeed, the proteins of the regulatory network catalyze chemical reactions, thereby controlling the links of the metabolic network. Similarly, frequent protein-gene interactions couple the protein interaction network with the genes and the DNA.

Page 184 The Watson quote came from James D. Watson, *Molecular Biology of the Gene*, 2nd ed. (New York: W. A. Benjamin, 1970): 99.

Page 185 WIT (What Is There?) is an integrated system for comparative analysis of sequenced genomes. What was important for us is that it also supports metabolic reconstruction from the sequence data. It can be reached through its Webpage found at http://www-unix.mcs.anl.gov/compbio/.

Page 185 Our results on the scale-free nature of the metabolism was published in H. Jeong, B. Tombor, R. Albert, Z. N. Oltvai, and A.-L. Barabási, "The large-scale organization of metabolic networks," *Nature* 407 (2000): 651–654.

Page 186 Andreas Wagner and David A. Fell's work on metabolism was published as "The Small World Inside Large Metabolic Networks," *Proceedings of the Royal Society, London*, B, 268 (2001): 1803–1810.

Page 187 Comparing the metabolic network of different organisms can shed light on the evolutionary relationships between different species as well. See J. Podáni, Z. N. Oltvai, H. Jeong, B. Tombor, A.-L. Barabási, and E. Szathmáry, "Comparable System-Level Organization of Archaea and Eukaryotes," *Nature Genetics* 29 (2001): 54–56; C. V. Forst and K. Schulten, "Phylogenetic Analysis of Metabolic Pathways," *Journal of Molecular Evolution* 52 (2001): 471–489.

Page 188 The yeast two-hybrid technique was developed in S. Fields and O. Song, "A Novel Genetic System to Detect Protein-Protein Interactions," *Nature* 340 (1989): 245–246. For recent developments in the technique, see also Li Zhu and Gregory J. Hannon, eds., *Yeast Hybrid Methods* (Natick, MA: Eaton, 2000).

Page 188 The comprehensive interaction map for yeast was published by P. Uetz, *et al.* in "A Comprehensive Analysis of Protein-Protein Interactions in *Saccharomyces cerevisiae*" *Nature* 403 (2000): 623–627; T. Ito *et al's* "Toward a Protein-Protein Interaction Map of the Budding Yeast: A Comprehensive System to Examine Two-Hybrid Interactions in All Possible Combinations Between the Yeast Proteins," *Proceedings of the National Academy of Sciences* 97 (2000): 1143–1147; and "A Comprehensive Two-Hybrid Analysis to Explore the Yeast Protein Interactome," *Proceedings of the National Academy of Sciences* 98 (2001): 4569–4574.

Page 188 The scale-free nature of the protein interaction networks in yeast is discussed in Hawoong Jeong, Sean Mason, Albert-László Barabási, and Zoltán N. Oltvai, "Centrality and Lethality of Protein Networks," *Nature* 411 (2001): 41–42. This paper contains the discussion of the relationships between lethality and topology as well. For a perspective on the results, see the News and Views article accompanying the paper J. Hasty and J. J. Collins, "Protein Interactions—Unspinning the Web," *Nature* 411 (2001): 30–31.

Page 189 For independent confirmation of power laws in the yeast protein network, as well as the potential link to gene duplication, see Andreas Wagner, "The Yeast Protein Interaction Network Evolves Rapidly and Contains Few Redundant and Duplicate Genes," *Molecular Biology and Evolution* 18 (2001): 1283–1292.

Page 189 For the results on protein domain networks, see Stefan Wuchty, "Scale-Free Behavior in Protein Domain Networks," *Molecular Biology and Evolution* 18 (2001): 1694–1702. For a different approach to the yeast protein network also demonstrating a scale-free structure see Jong Park, Michael Lappe, and Sarah A. Teichmann, "Mapping

Protein Family Interactions: Intramolecular and Intermolecular Protein Family Interaction Repertoires in the PBD and Yeast," *Journal of Molecular Biology* 307 (2001): 929–938. For results on the *H. pylori* protein network, see Hawoong Jeong, Sean Mason, Albert-László Barabási, and Zoltán N. Oltvai, "Centrality and Lethality of Protein Networks," *Nature* 411 (2001): 41–42.

Page 190 On gene duplication and its evolutionary role, see John Maynard Smith and Eörs Szathmáry, *The Origins of Life* (Oxford, England: Oxford University Press, 1999).

Page 190 The papers that independently suggested gene duplication as the source of the scale-free topology in regulatory networks are A. Bhan, D. J. Galas, and T. G. Dewey, "A Gene Duplication Growth Model of Scaling in Gene Expression Networks (to be published); A. Vasquez, A. Flammini, A. Maritan, and A. Vespignani, *Modeling of Protein Interaction Networks* (http://xxx.lanl.gov/abs/cond-mat/0108043); and R. V. Solé, R. Pastor-Satorras, E. D. Smith, and T. Kepler, "A Mode of Large-Scale Proteome Evolution (Santa Fe Preprint, available at www.santafe.edu 2001). See also J. Giam, N.M Luscombe, and M. Gerstein, "Protein Family and Fold Occurrences in Genomes: Power-law Behavior and Evolutionary Model," Journal of Molecular Biology, 313 (2001): 673–681. Note that the gene duplication models have a number of fascinating properties from the perspective of network theory. For a more detailed discussion see J. Kim, P.L. Krapivsky, B. Kahng, and S. Redner, "Evolving Protein in Interactive Networks," http://xxx.lanl.gov/abs/condmat/0203167.

Page 192 Lane, Levine, and Vogelstein have been awarded all of the possible honors and prizes in medicine, leading many to believe that it is just a matter of time before they receive the Nobel as well. Indeed, David Lane, currently one of the top cited scientists in the UK, was knighted Sir David Lane by Queen Elizabeth in 2000. Arnold J. Levine, currently the president of the prestigious Rockefeller University in New York, was the first recipient of the Albany Medical Center Prize, which at $500,000 is second among medical awards only to the Nobel prize in value. Vogelstein, currently a Howard Hughes Investigator at Johns Hopkins School of Medicine, continues to produce an unparalleled string of significant discoveries, three of his publications now ranking among the ten most cited papers in medicine.

Page 192 For the suggestion that networks play a key role in understanding cancer, see Bert Vogelstein, David Lane, and Arnold J. Levine, "Surfing the p53 Network," *Nature* 408 (2000): 307–310. Note that this paper does not perform a quantitative analysis of the network; it offers rather compelling empirical arguments on the scale-free nature of the networks. We have subsequently analyzed the network's topology, finding that, indeed, with a good approximation, it is scale-free (Hawoong Jeong, D. A. Mongru, Z. N. Oltvai, and A.-L. Barabási, unpublished).

Page 194 The microarray technology, introduced in 1991 by Stephen Fodor and his collaborators (see S. P. A. Fodor, J. L. Read, M. C. Pirrung, L. Stryer, A. T. Lu, and D. Solas, "Light-Directed, Spatially Addressable Parallel Chemical Synthesis," *Science* 251 [1991]: 767–773), allows researchers to decipher the dynamics of gene interactions within the cell. The consequences of this breakthrough have already fundamentally altered the way biology is done in most laboratories. It is only a question of time before everything from diagnostics in the doctor's office to drug development is changed. A *DNA chip*, or *microarray*, is a silicon or glass wafer patterned with a technology used by computer chip makers. Photolithography machines etch an array of tiny holes just barely large enough to allow robotic arms to place in these holes short DNA strands, each hole containing a different gene. Thus with 30,000 holes you can place a copy of each gene of the human genome on a single chip. When the DNA in a cell generates proteins, a gene is first copied into a unique messenger RNA (mRNA) molecule, which later is translated into a protein. Thus the number and type of mRNA molecules in a cell closely track the orders given by the DNA. Each mRNA molecule can stick to only one microarray hole, the one that contains the matching piece of DNA that produced the mRNA molecule in the first place. If a biologist studying a rare disease places a culture of the ill cells on the DNA chip, the holes corresponding to the active genes will fill up with mRNA strands, while all other holes will stay empty. A laser reader will scan through each hole, indicating which genes are busy producing proteins. Therefore, the measurement will tell you which genes function normally and which are those that are shut down by some genetic disorder.

Page 195 For a very general discussion of the effect of new biological tools—such as microarrays—on the future of medicine and drug development, see the special issue of *Time* magazine of January 15, 2001, titled "Drugs of the Future."

Page 195 A wonderful demonstration of the microarray's ability to follow genes being switched on and off has been offered by several recent papers that identified groups of genes being simultaneously active during different stages of the cell cycle. See Neal S. Holter, Madhusmita Mitra, Amos Maritan, Marek Cieplak, Jayanth R. Banavar, and Nina V. Fedoroff, "Fundamental Patterns Underlying Gene Expression Profiles: Simplicity from Complexity," *Proceedings of the National Academy of Sciences* 97 (2000): 8409–8414; and Orly Alter, Patrick O. Brown, and David Botstein, "Singular Value Decomposition for Genome-Wide Expression Data Processing and Modeling," *Proceedings of the National Academy of Sciences* 97 (2000): 10101–10106.

Page 196 Note that, though there are only about 30,000 or so genes in the human genome, the number of proteins can be much higher. This is due to a process called *alternate splicing*, in which the messenger RNA is chopped up and reattached in many different ways, creating different proteins. Therefore, in eukaryotes the number of proteins is much larger than the number of genes, defying the one gene–one protein dogma of molecular biology, valid in bacterium.

Page 197 For a discussion of the genome's complexity and the role of the genetic network, see Jean-Michel Claverie, "What If There Are Only 30,000 Human Genes?" *Science* 291 (2001): 1255–1257.

THE FOURTEENTH LINK: NETWORK ECONOMY

Page 199 For a general discussion of the role of networks in business and economy, see E. Bonabeau, *The Alchemy of Networks: Network Science Applied to Business* (in preparation).

Page 199 The Time-Warner and AOL merger is described in detail in Daniel Okrent, "Happily Ever After?" *Time*, January 24, 2000. See also the extensive joint interview with AOL's Steve Case and Time-Warner's Jerry Levin in the same issue of the magazine.

Page 200 For the story of the Daimler-Benz and Chrysler merger, see Bill Vlasic and Bradley A Stertz, *Taken for a Ride: How Daimler-Benz Drove Off with Chrysler* (New York: William Morrow, 2000).

Page 200 Not all mergers are the result of an expanding market and economy—the first wave of mergers in fact began after the worldwide depression of 1883. For a brief historical perspective on mergers see David Besanko, David Dranove, and Mark Shanley, *Economics of Strategy* (New York: John Wiley, 2000): 198–199.

Page 201 The hierarchical organization has a history of over a century. "In our scheme, we do not ask for the initiative of our men. . . . All we want them is to obey orders we give them, do what we say, and do it quick." Thus wrote Frederick Winslow Taylor, the father of scientific management, at the beginning of the twentieth century, summarizing a philosophy that is responsible for the wealth and material culture as we know it. As Brink Lindsey writes in *The Man with the Plan* (Reason Online, 1998, www.reason.com) before Taylor, manufacturing followed a craft system. The secrets of the craft were well kept and passed down, grudgingly, from master to apprentice. The real work potential of the shop was jealously guarded by the craftsmen because the compensation system was based on the number of pieces produced and not by hourly wages. Taylor, using a stopwatch, singlehandedly changed that. He broke down all manufacturing processes into simple elements, standardizing the work tasks and compensation process. His groundbreaking approach to manufacturing turned Bethlehem Steel into the world's most modern factory, reducing yard workers' ranks from 500 to 140 while doubling production. A victim of his own success, Taylor was eventually fired as a result of anger over the layoffs. Yet after him no factory could compete without fully adopting his methods. Taylor introduced a clear separation of ranks, reducing workers to drone-order executors. He invented the white-collar worker, responsible for planning every single manufacturing step and making sure that the workers faithfully execute them. His most important legacy is what we call the vertical organization, which set in stone for a full century the structure of the web within the firm. The life of and work of Frederick W. Taylor has been the subject of several biographies and scientific monographs. See Robert Kanigel, *The One Best Way: Frederick Winslow Taylor and the Enigma of Efficiency* (New York: Viking Penguin, 1997). For a standard biography see Frank Barkley Copley, *Frederick W. Taylor, Father of Scientific Management*, 2 vols. (New York: Taylor Society, 1923). Taylor's own influential work is *The Principles of Scientific Management* (New York: Harper & Brothers, 1915: reprint, Mineola, N.Y.: Dover, 1998).

Page 201 Note that Ford's factories also played a key role in the birth of modern manufacturing. It is there the moving assembly line was developed, a key component of all mass-production plants. For a short summary of the history of the developments at Ford and the players behind the development of the assembly line, see Joseph B. White, *The Line Starts Here* (www.wsj.com/public/current/articles/SB915733342173968000.htm, Wall Street Journal Interactive, 2000).

Page 201 Classical economic theory views organizations, firms, and corporations as optimized networks aiming to achieve the largest financial output with the fewest resources. This is Taylor's legacy, maintaining that running a company is an optimization process aimed at increasing profits. Such profit-driven optimization favors a tree structure. Indeed, if manufacturing is the company's primary goal, costs can be significantly reduced by assigning all repetitive, specialized tasks to low-income workers. Recent studies indicate that information is also most efficiently managed within a hierarchical organization, since a tree avoids unnecessary duplication of information and communication. As each firm's activity is a combination of manufacturing and information management, the pyramid structure appears to be here to stay. For a detailed discussion of the hierarchical tree within the firm, see Patrick Bolton and Mathias Dewatriport, "The Firm as a Communication Network," *Quarterly Journal of Economics* 109 (Nov. 1994): 809–839.

Page 202 For a general discussion on the shift within the firm toward a network organization, see *Business Week*'s special double issue "The 21st Century Corporation," August 21–28, 2001.

Page 202 For a concise review of network theories for organizations, see Peter R. Monge and Noshir S. Contractor, "Emergence of Communication Networks," in *The New Handbook of Organizational Communication*, ed. Fredric M. Jablin and Linda L. Putnam (Thousand Oaks, Calif.: Sage Publications, 2001): 440–502.

Page 203 For Jordan's role in the Clinton-Lewinsky scandal, see Eric Pooley, "The Master Fixer Is a Fix," *Time*, Feb. 2, 1998.

Page 203 For a more regional example of interlocked directorships in the Chicago area, see Melissa Allison, "Directors Weave a Complex Web," *Chicago Tribune*, June 17, 2001, sec, 5, p. 1–2.

Page 204 For a detailed discussion of the corporate network, see Gerald F. Davis, Mina Yoo, and Wayne E. Baker's "The Small World of the Corporate Elite" (Preprint, February 2001).

Page 204 For a mathematical analysis of the director's network, see M. E. J. Newman, S. H. Strogatz, and D. J. Watts, "Random Graphs with Arbitrary Degree Distributions and Their Applications" *Physical Review*, E 64 (2001): 026118.

Page 205 For Jordan's path in the corporate world, see chapter 12 in Vernon E. Jordan, Jr.'s autobiography, with Annette Gordon-Reed, *Vernon Can Read (A Memoir)* (New York: Public Affairs, 2001) and the above cited work of Davis, Yoo and Baker.

Page 206 For the power of networks in Silicon Valley, see Emilio J. Castilla, Hokyo Hwang, Ellen Granovetter, and Mark Granovetter, "Social Networks in Silicon Valley," in *The Silicon Valley Edge: A Habitat for Innovation and Entrepreneurship*, ed. Chong-Moon Lee, William F. Miller, Marguerite Gong Hancock, and Henry S. Rowen (Cambridge, England: Cambridge University Press, 2001), 218–247.

Page 207 Walter W. Powell, Douglas White, and Kenneth W. Koput, "Dynamics and Movies of Social Networks in the Field of Biotechnology: Emergent Social Structure and Process Analyses," (preprint, April 12, 2001).

Page 207 For detailed mathematical analysis of the networks behind the pharmaceutical industry see M. Riccaboni, F. Pammolli, and G. Caldarelli, "Complexity of Connections in Social and Economical Structures" (preprint, 2001).

Page 208 For an another example of small worlds in the economy, see Bruce Kogut and Gordon Walker, "The Small World of Germany and the Durability of National Networks," *American Sociological Review* 66 (2001): 317–335.

Page 208 Besides being scale-free, a network economy displays clustering as well. There is first a strong geography-based clustering, where companies have more links to local consumers. *Globalization*, the buzzword of the last decade, actually means the proliferation of long-range, geography-defying links—companies finding consumers and vendors not locally but worldwide. Then there is industry-based clustering—companies in the same market or business sharing many links. Such links, though geographically biased, cut easily across large distances. The clustered nature of the economy has been documented recently in Bruce Kogut and Gordon Walker, "The Small World of Germany and the Durability of National Networks." *American Sociological Review*, 66 (2001): 317–335. They investigate firm ownership in Germany, mapping out the links between five hundred nonfinancial corporations, twenty-five banks, and twenty-five insurance firms. In this network two firms are connected if they have a common owner. The obtained network in a way is rather similar to the actor network, where the actors correspond to companies, and movies to owners. A typical owner owns multiple companies, just as there are many actors in a single movie. The analysis of the obtained company network clearly indicates that German firms are part of a small world. The diameter of the network is 4.81, i.e., the majority of these companies are linked though a chain of four owners. Kogut and Walker have found a huge clustering coefficient as well. If the companies were to form a random network, the chance of finding a link between two neighbors of a certain firm is expected to be 0.5 percent. In contrast, in the real network two neighbors of any firm have a 67 percent chance of having a common owner. This is clearly a significant difference underlying the very high degree of clustering characterizing the economy.

Page 209 Walter W. Powell, "Inter-Organizational Collaboration in the Biotechnology Industry," *Journal of Institutional and Theoretical Economics* 512 (1996): 197–215.

Page 209 One of the pioneers of the idea that the economy should be viewed as an evolving network is Alan Kirman of University of Aix-Marseille. His papers offer a truly excellent discussion of the shortcomings of current economic thinking and the role of networks in economic theories. See "The Economy as an Evolving Network," *Journal of Evolutionary Economics* 7 (1997): 339–353; "Aggregate Activity and Economic Organization," *Revue europeenne des sciences sociales* 37, no. 113 (1999): 189–230; and "The Economy as an Interactive System," in *The Economy as an Evolving Complex System II* (Proceedings of the Santa Fe Institute Studies in the Sciences of Complexity, vol. 27), ed. W. Brian Arthur, Steven N. Durlauf, and David A. Lane (Reading, MA, Addison-Wesley, 1997), 491-532.

Page 209 The Asian economic crisis was widely documented in the press and in scholarly articles alike. For a day-to-day breakdown of the events, see the Website maintained by Nouriel Roubini, associate professor of economics and international business at the Stern School of Business of New York University. The site, titled "Chronology of the Asian Currency Crisis and Its Global Contagion," is available at http://www.stern.nyu.edu/~nroubini/asia/AsiaChronology1.html. For a discussion of the origins of the crisis, see Giancarlo Corsetti, Paolo Pesenti, and Nouriel Roubini, "What Caused the Asian Currency and Financial Crisis?" *Japan and the World Economy*, Sept. 1999, 305-373.

Page 210 *Economic Report of the President* (Washington, D.C.: U.S. Government Printing Office, 1999).

Page 210 Paul Krugman, *What Happened to Asia?* (January 1998) http://web.mit.edu/krugman/www/DISINTER.html.

Page 212 For a detailed discussion of the effects of outsourcing and the story of Cisco, Compaq, and other apostles of the network economy, see Bill Lakenan, Darren Boyd, and Ed Frey, "Why Cisco Fell: Outsourcing and Its Perils," *Strategy + Business* (3rd quarter 2001): 54–65.

Page 213 The Hotmail story is described by Steve Jurvetson, the partner in the venture capital firm that secured the seed funding for the company, in "Turning Customers into Sales Force," *Business 2.0*, Nov. 1, 1998. See also the portrait of Sabeer Bhatia in Stuart Whitmore "Driving Ambition," Asiaweek.com, http://www.asiaweek.com/asiaweek/technology/990625/bhatia.html.

Page 216 Note that there is increasing interest in academic literature on economic networks. For some representative examples of the work in this area, see Matthew O. Jackson and Alison Watts, "The Evolution of Social and Economic Networks," *Journal of Economic Literature* (in press, 2001); Alison Watts, "A Dynamic Model of Network Formation," *Games and Economic Behavior* 34 (2001): 331–341; Matthew O. Jackson and Alison Watts, "On the Formation of Interaction Networks in Social Coordination Games," *Journal of Economic Literature* (in press, 2001); Venkatesh Bala and Sanjeev Goyal, "A Noncooperative Model of Network Formation," *Econometrica* 68 (2000): 1181–1229; "Learning, Network Formation and Coordination" (preprint); and "A Strategic Analysis of Network Reliability," *Review of Economic Design* 5 (2000): 205–228; Nigel Gilbert, Andreas Pyka, and Petra Ahrweiler, "Innovation Networks: A Simulation Approach," *Journal of Artificial Societies and Social Simulation* 4, no. 3 (2001); Lawrence E. Blume and Steven N. Durlauf, *The Interactions-Based Approach to Socioeconomic Behavior*, http://www.ssc.wisc.edu/econ/archive/wp2001.htm; Nicholas Economides, "Desirability of Compatibility in the Absence of Network Externalities," *American Economic Review* 78 (1989): 108–121; "Compatibility and the Creation of Shared Networks," in *Electronic Services Networks: A Business and Public Policy Challenge*, ed. Margaret Guerin-Calvert and Steven Wildman (New York: Praeger, 1991); "Network Economics with Application to Finance," *Financial Markets, Institutions & Instruments* 2 (1993): 89–97; Nicholas Economides and Steven C. Salop, "Competition and Integration Among Complements and Network Market Structure," *Journal of Industrial Economics* 40, no. 1 (1992): 105–123. See also D. McFadzean, D. Stewart, and L. Tesfatsion, "A Computational Laboratory for Evolutionary Trade Networks," *IEEE Transactions on Evolutionary Computation* 5 (2001): 546–560; L. Tesfatsion, "A Trade Network Game with Endogenous Partner Selection," in *Computational Approaches to Economic Problems*, ed. H. M. Amman, B. Rustem, and A. B. Whinston (Kluwer Academic, 1997), 249–269. See also the Websites of Leigh Tesfatsion, http://www.econ.iastate.edu/tesfatsi/netgroup.htm, and Nicholas Economides, http://www.stern.nyu.edu/networks/site.html, with numerous links to researchers and papers focusing on network economics.

Page 216 Note that a rapidly expanding field within physics aims to address economic phenomena in quantitative terms, using the tools of statistical mechanics. For a short introduction, see Rosario N. Mantegna and H. Eugene Stanley, *An Introduction to Econophysics: Correlations and Complexity in Finance* (Cambridge, England: Cambridge University Press, 2000); Jean-Phillipe Bouchaud, Marc Potters, *Theory of Financial Risk: From Statistical Physics to Risk Management* (Cambridge, England: Cambridge University Press, 2000). See also J. Doyne Farmer, "Physicists Attempt to Scale the Ivory Towers of Finance," *IEEE Computing in Science and Engineering* (Nov.–Dec. 1999): 26–39. Most of this network focuses on the properties of stock fluctuations. For the relationship between networks and the stock market, see Hyun-Joo Kim, Youngki bee, Im-mook Kim, and Byungnam Kahng, "Scale-Free Networks in Financial Correlations," http://xxx.lanl.gov/abs/cond-mat/0107449.

Page 216 Many companies are experimenting with the incorporation of network ideas under various business models. For example, Ecrush.com asks you to tell them if you have a crush on somebody. They will send your crush a secret message saying "someone likes you" and invite them to sign up, too. If your crush signs up and lists you as a crush, the program matches you up. If your crush does not list you as a crush, they can never find out who it was that approached her or him. ICQ.com, another network-obsessed startup that boasts a whopping 116 million users, is less ambitious and more down to earth. It offers you an environment to activate your links efficiently. Free software monitors your list of friends, telling you which of them is online, offering you the possibility to reach them instantly.

Page 216 For a discussion of interactions between economic institutions and policy making, see P. Cooke and K. Morgan, "The Networks Paradigm: New Departures in Corporate and Regional Development," *Environment and Planning, D: Society and Space* 11 (1993): 543–564.

Page 217 For a discussion on policy networks, see David Marsh, ed., *Comparing Policy Networks* (Buckingham: Open University Press, 1998); Dirk Messner, *The Network Society* (London: Frank Cass, 1997); and Manuel Castell, *The Rise of the Network Society* (London: Blackwell, 1996).

The Last Link: Web Without a Spider

Page 222 For a discussion of the network behind the terrorist cell responsible for the September 11 attack, see www.orgnet.com, Valdis Kreb's Website. See also Thomas A. Steward, "Six Degrees of Mohamed Atta," *Business 2.0*, Dec. 2001, 63.

Page 223 For a discussion of fighting a network organization in a netwar, see John Arquilla and David F. Ronfeldt, eds., *Networks and Netwars* (Santa Monica, CA: RAND Corp., 2001); and Thomas A. Steward, "Americas' Secret Weapon," *Business 2.0*, Dec. 2001, 58–68.

Page 224 The work of Christo and Jean-Claude is the subject of many books and monographs. See Jacob Baal-Teshuva, *Christo and Jeanne-Claude* (Cologne, Germany: Taschen, 2001). The "revelation through concealment" phrase comes from David Bourdon, *Christo* (New York: Abrahams, 1970).

Page 225 Note that complexity is a huge subject, and many researchers from physicists to mathematicians and biologists are working on various ways to approach it. For an array of books covering different approaches, see notes to Chapter 1.

Afterlink:-Hierarchies and Communities

Page 228 For a recent review on the applications of networks to biology, see the *Nature Insight on Computational Biology*, a collection of articles published in *Nature* 420 (2002): 205–251. For search in complex information networks, like the World Wide Web, see Lada A. Adamic, Rajan M. Lukose, and Bernardo A. Huberman, "Local Search in Unstructured Networks," in *Handbook of Graphs and Networks: From the Genome to the Internet*, edited by S. Bornholdt and H. G. Schuster (Berlin: Wiley-VCH, 2002); L. A. Adamic, R. M. Lukose, A. R. Puniyani, and B. A. Huberman, "Search in Power-Law Networks," *Physical Review*, E 64 (2001): 046135. For research on the Gnutella network, see Matei Ripeanu, Ian Foster, Adriana Iamnitchi, "Mapping the Gnutella Network: Properties of Large-Scale Peer-to-Peer Systems and Implications for System Design," *IEEE Internet Computing Journal* 6 (2002): 50–57. For the network structure of the Marvel Universe, the network of comics characters, see R. Alberich, J. Miro-Julia, and F. Rossello, "Marvel Universe Looks Almost Like a Real Social Network," http://xxx.lanl.gov/abs/cond-mat/0202174.

Page 228 The immunologist quoted in the text is H.-G. Thiele, the former director of the department of immunology of the University of Hamburg, Germany. See page 228 in H.-G. Thiele, "Contemplations on the Paradigm of Self and Nonself Discrimination and on Other Concepts Ruling Contemporary Immunology," *Cellular and Molecular Biology* 48 (2002): 221–236.

Page 229 For a recent update on network research, see *Handbook of Graphs and Networks: From the Genome to the Internet,* edited by S. Bornholdt and H. G. Schuster (Berlin: Wiley-VCH, 2002); S. N. Dorogovtsev and J. F. F. Mendes, *Evolution of Networks: From Biological Nets to the Internet and WWW* (Oxford: Oxford University Press, 2003).

Page 229 For the CNN coverage on multitasking, see Porter Anderson, "Study: Multitasking Is Counterproductive (Your Boss May Not Like This One)," http://www.cnn.com/2001/CAREER/trends/08/05/multitasking.study/. For the original research, see Joshua S. Rubinstein, David E. Meyer, and Jeffrey E. Evans, "Executive Control of Cognitive Processes in Task Switching," *Journal of Experimental Psychology: Human Perception and Performance* 27 (2001): 763–797.

Page 231 For *Nature*'s coverage on the perils of scientific forecasting, see Philip Cambell, "Tales of the Expected," *Nature* 402 (1999): C7–C9; J. L. Heilbron and W. F. Bynum, "Plus çà change," *Nature* 402 (1999): C86–C88.

Page 231 For the modular hypothesis, see Leland H. Hartwell, Andrew W. Murray, John J. Hopfield, and Stanislas Leibler, "From Molecular to Modular Cell Biology," *Nature* 402 (1999): C47–C52. See also D. A. Lauffenburger, "Cell Signaling Pathways as Control Modules: Complexity for Simplicity?" *Proceedings of the National Academy of Sciences USA* 97 (2000): 5031–5033. C. V. Rao and A. P. Arkin, "Control Motifs for Intracellular Regulatory Networks," *Annual Review of Biomedical Engineering* 3 (2001): 391–419.

Page 234 The toy model described had a quite adventurous history. In 2000, Tamás Vicsek, my former thesis advisor from Eötvös University, Budapest, give a talk at the University of Notre Dame about his ongoing research. By then my group was fully infected by networks, which prompted Tamás to ask us a straightforward question: Could we build a deterministic network model, one that would generate a scale-free network that would have a quite fixed, nonrandom architecture? Fractals, the research field in which Tamás is one of the world's experts, has benefited significantly from such visually and computationally simple and appealing models. Yet most of the network models we could come up for several days did not work—we were generating mostly networks without hubs. Tamás, however, on the plane back to Hungary did come up with a construction. Immersed in several other projects, we all forgot about it and moved on. During the summer of 2001, motivated by our desire to understand modularity, Erzsébet Ravasz and I again took up the project, and, having all but forgotten about Tamás's successful solution, we designed a deterministic scale-free network. A few days later I flew to Budapest, and during a meeting with Tamás, we realized that the model Erzsébet and I designed was very close to the one Tamás had come up with a year earlier. The final publication, first discussing deterministic modular scale-free networks, came out a few months later: Albert-László Barabási, Erzsébet Ravasz, and Tamás Vicsek, "Deterministic Scale-Free Networks," *Physica* A 299 (2001): 559–564. The model described in this chapter is a version developed a bit later, published in E. Ravasz, A. L. Somera, D. A. Mongru, Z. N. Oltvai, and A.-L. Barabási, "Hierarchical Organization of Modularity in Metabolic Networks," *Science* 297 (2002): 1551–1555. For a more detailed investigation of the model and the presence of hierarchy in real networks, see Erzsébet Ravasz, Albert-László Barabási, "Hierarchical Organization in Complex Networks," *Physical Review*, E (in press), http://xxx.lanl.gov/abs/cond-mat/0206130.

Page 235 The Porto group's paper on the scaling of the clustering coefficient in hierarchical networks was published in S. N. Dorogovtsev, A. V. Goltsev, and J. F. F. Mendes, "Pseudofractal Scale-Free Web," *Physical Review*, E 65 (2002): 066122.

Page 236 For hierachical organization in biological networks see E. Ravasz, A. L. Somera, D. A. Mongru, Z. N. Oltvai, and A.-L. Barabási, "Hierarchical Organization of Modularity in Metabolic Networks," *Science* 297 (2002): 1551–1555. Hierarchy in other systems is discussed in Erzsébet Ravasz and Albert-László Barabási, "Hierarchical Organization in Complex Networks," *Physical Review*, E (in press) http://xxx.lanl.gov/abs/cond-mat/0206130. For evidence of the scaling of the clustering coefficient for the World Wide Web, see Jean-Pierre Eckmann and Elisha Moses, "Curvature of Co-Links Uncovers Hidden Thematic Layers in the World Wide Web," *Proceedings of the National Academy of Science USA* 99 (2002): 5825–5829. For hierarchy in the Internet, see A. Vazquez, R. Pastor-Satorras, and A. Vespignani, "Large-Scale Topological and Dynamical Properties of the Internet," *Physical Review*, E 65 (2002): 066130.

Index